Lectures in Mathematics
ETH Zürich
Department of Mathematics
Research Institute of Mathematics

Jon F. Carlson

Modules and Group Algebras

Notes by Ruedi Suter

Birkhäuser Verlag
Basel · Boston · Berlin

Author's address:

Jon F. Carlson
Department of Mathematics
University of Georgia
Athens, Ga 30602
USA

Mathematics Subject Classification (1991): 20C05, 20C20

A CIP catalogue record for this book is available from the Library of Congress, Washington D.C., USA

Deutsche Bibliothek Cataloging-in-Publication Data
Carlson, Jon:
Modules and group algebras / Jon F. Carlson. Notes by Ruedi Suter. – Basel ; Boston ; Berlin : Birkhäuser, 1996
(Lectures in mathematics ETH Zürich)
ISBN-13: 978-3-7643-5389-6 e-ISBN-13: 978-3-0348-9189-9
DOI: 10.1007/978-3-0348-9189-9

Printed on acid-free paper produced of chlorine-free pulp. TCF ∞

ISBN-13: 978-3-7643-5389-6

9 8 7 6 5 4 3 2 1

To Tina

Preface

The notes in this volume were written as a part of a Nachdiplom course that I gave at the ETH in the summer semester of 1995. The aim of my lectures was the development of some of the basics of the interaction of homological algebra, or more specifically the cohomology of groups, and modular representation theory. Every time that I had given such a course in the past fifteen years, the choice of the material and the order of presentation of the results have followed more or less the same basic pattern. Such a course began with the fundamentals of group cohomology, and then investigated the structure of cohomology rings, and their maximal ideal spectra. Then the variety of a module was defined and related to actual module structure through the rank variety. Applications followed. The standard approach was used in my University of Essen Lecture Notes [C1] in 1984. Evens [E] and Benson [B2] have written it up in much clearer detail and included it as part of their books on the subject.

In the last three years there have been several advances which suggest an entirely new approach to the subject. Basically the shift has been towards a much more categorical view of representation theory, and an expansion of the viewpoint to include infinitely generated modules as well as the finitely generated ones. The real surprise has been that some of the constructions in the category of all modules have had new and original applications for the category of finitely generated modules. All of this is recent. At the time of this writing only a few of the works have appeared in print.

Modular representation theory as a subject had its origins in the work of Dickson and others around the early part of the century. However the beginnings were fairly modest and it fell to Richard Brauer to bring the area to maturity during the 1940's and 1950's. Almost single-handedly, Brauer developed most of the now standard character theory and block theory for modular group algebras. The block theory continues to be a subject of very active research. In the last decade it has found new vigor in the challenge of the conjectures of Alperin and Broué.

The mid to late 1970's saw the emergence of a parallel theory, concerned with the structure of kG-modules (k a field of characteristic $p > 0$, G a finite group) in general rather than just the block distributions of the irreducible and projective modules. There is a legendary story that Brauer, himself, used to advise his students not to try to study the representation theory of p-groups. The subject seemed to be too difficult with little or no promise of productive results. Yet for the investigation of module structure, many of the most difficult and fascinating problems can be easily reduced to questions involving the representations of p-groups over fields of characteristic p. On the other hand, in this situation, all group characters are trivial, the Grothendieck group is trivial and many of the classical techniques of representation theory have no relevance. The only method left open to us is homological algebra.

The foundations of a module theory for modular group algebras were laid in the Evens-Venkov proof that the cohomology ring $H^*(G,k)$ is finitely generated as a k-algebra, and Quillen's characterization of the component of the maximal ideal spectrum of $H^*(G,k)$. Chouinard's Theorem, that a kG-module is projective if and only if its restriction to every elementary abelian p-subgroup is projective, had much the same flavor as the work of Quillen. In the late 1970's Alperin and Evens generalized first Chouinard's Theorem, using Alperin's notion of complexity, and then generalized Quillen's result for support varieties of finitely generated modules. The last was also done independently by Avrunin. Numerous advances have been made since then. See the aforementioned books [B2, E] for a more complete account.

In the last few years, the language and concepts of category theory have become standard in many parts of representation theory. Some of the most intense investigations in modern block theory have centered on problems of the existence of stable or derived equivalences of blocks. For the module theory the expansion of viewpoint has focused on the nature of other cohomology theories. In essence this has meant looking at quotients by thick subcategories of the stable category of modules modulo projectives. A beginning was made in [CDW] in the characterization of homomorphisms in the difference one complexity quotients. While the project settled a few questions it opened many more. One of the observations of [CDW] was that complexity quotient categories have no Krull-Schmidt Theorem, no uniqueness of decompositions of direct sums of objects. Later Rickard observed that the Krull-Schmidt property could be recovered using homotopy colimits if only infinite direct sums were allowed. Suddenly a rationale for looking at infinitely generated modules was created. However it meant that in order to use homotopy colimits and other methods borrowed from homotopy theory, it would be necessary to extend all of the notions of varieties and complexity into the category of all kG-modules. This has now been shown to work (see [BCR1, BCR2]).

For me, one of the most satisfying parts of this is that the translation from the category of finitely generated modules to the category of all modules was not simply a matter of slavishly adapting theorems from one setting to another. Instead it has required and led to the discovery of some genuinely new structures. The most striking of these is Rickard's idempotent modules in the stable category [R]. These are modules M with the property that $M \otimes_k M \cong M \oplus P$ for some projective kG-module P. It can be shown that the trivial module is the only non-projective finitely generated module with this property. However in the category of all kG-modules such idempotents exist in pairs associated to any thick subcategory which is ideally closed under tensor products. Under the right conditions the idempotents can be used to classify the thick subcategories of the stable category. In another remarkable application, Benson has used all of this machinery to settle some questions on the varieties of modules M with $H^*(G,M) = 0$. The final section of these notes sketches another application which is an extension of an observation of Benson.

These notes are not intended to be an encyclopedic reference to the new results. They were notes written for a graduate course and are really only meant to be an introduction to the concepts necessary to understand the recent advances. The first seven sections contain mostly old material. They cover the basics of modules over modular group algebras, diagrams for modules, triangulated categories, the fundamentals of cohomology and the numerous incarnations of the cohomology (cup) products. Sections 8 and 9 concern a relative homological algebra and applications showing that ideals in the cohomology ring can be represented by exact sequences in the module category. Most of this comes from joint work with Wayne Wheeler and Chuang Peng. Section 10 contains a brief introduction to varieties for modules while Section 11 is a preparation for the study of infinitely generated modules. In Section 12 we use some of the earlier results to construct the idempotent modules corresponding to the thick subcategory of the stable category consisting of all finitely generated kG-modules whose varieties are contained in a fixed closed subset of the maximal ideal spectrum of $H^*(G, k)$. In the final section we sketch an application of the technology.

If this had been a two semester course then I might have attempted to get into some of the details of the work on varieties of infinitely generated modules in [BCR1] and [BCR2]. As it was, even the development of the idempotent modules was only presented for a special case. Still, I hope the reader will find the notes to be a useful introduction to the ideas.

Finally, I would like to thank all of the people who made these notes possible and made my stay in Zürich so enjoyable and productive. Particularly Ruth Ebel and Rahel Boller of the ETH were most helpful with arranging all of the little things that are so necessary to make a long stay away from home seem civilized. Urs Stammbach and Guido Mislin both suffered through my lectures and provided lots of informative comments as well as corrections to my numerous errors. Most especially I need to thank Ruedi Suter for his hard work and for the excellent job that he has done on these notes. I can say with confidence that any deficiency in the manuscript is mine alone.

September, 1995 Jon F. Carlson

CONTENTS

Notation

Throughout these notes G is a finite group, and k is a field of positive characteristic p. We write Hom and $\otimes$ for the functors that one usually denotes by Hom_k and $\otimes_k$, respectively. Let ${}_{kG}\mathfrak{mod}$ denote the category of finitely generated left kG-modules.

1 Augmentations, nilpotent ideals, and semisimplicity

In the first section we explore a few of the most fundamental properties of group algebras and their modules. One of our main points is the proof of Maschke's Theorem, which tells us that if the characteristic of k does not divide the order of G, then every exact sequence of kG-modules splits and every kG-module is both projective and injective. Therefore the application of homological algebra is interesting only in the complementary case that the characteristic of the coefficient field k divides the order of the group G. We begin with an assortment of loosely related facts.

We have the functor (forgetful functor)

$$_{kG}\mathfrak{mod} \longrightarrow {}_k\mathfrak{mod} = {}_k\mathfrak{vec}$$

to the category of finite-dimensional k-vector spaces which to every finitely generated kG-module associates its underlying k-vector space (which is finite-dimensional because $|G| < \infty$). So every module in $_{kG}\mathfrak{mod}$ casts its shadow into the realm of linear algebra. Indeed, if M is in $_{kG}\mathfrak{mod}$, then the elements of the group G act by k-linear transformations on the underlying vector space of M. So we have a homomorphism $\rho : G \to \mathrm{GL}_{\dim M}(k)$ which associates each element of G with the matrix of its action on M, relative to some chosen k-basis of M. The homomorphism ρ is known as the representation associated to the module M.

A basic fact about group algebras is that the Krull-Schmidt Theorem holds for kG, as for any finite-dimensional algebra. We refer the reader to [CR1, §6B] for a proof.

Theorem 1.1 (Krull-Schmidt) *Let M be a module in $_{kG}\mathfrak{mod}$. Then M is a direct sum $M = M_1 \oplus \cdots \oplus M_m$ of indecomposable kG-modules $M_1, \ldots, M_m$. Moreover, if $M = M_1' \oplus \cdots \oplus M_{m'}'$ and $M_1', \ldots, M_{m'}'$ are indecomposable, then $m' = m$ and there is a permutation $\pi \in S_m$ such that $M_i \cong M_{\pi(i)}'$ for $i = 1, \ldots, m$.*

We have the functor

$$S : {}_{kG}\mathfrak{mod} \longrightarrow \mathfrak{mod}_{kG}$$

to the category of finitely generated right kG-modules given by $S(M) = M$ as k-vector spaces for M in $_{kG}\mathfrak{mod}$, and $mg = g^{-1}m$ for $m \in S(M)$ and $g \in G$. For morphisms $M \xrightarrow{\alpha} N$ we have $S(\alpha)(m) = \alpha(m)$ for $m \in S(M) = M$. The functor S is an equivalence of categories. This explains why without loss of generality it suffices to consider left kG-modules only.

The augmentation map $\varepsilon = \varepsilon_G : kG \to k$ is the homomorphism of rings with 1, given by $\sum\limits_{g\in G} a_g\, g \mapsto \sum\limits_{g\in G} a_g$. It makes k a kG-module by defining $a \cdot 1 = \varepsilon(a)1$

for $a \in kG$. Since $g \cdot 1 = 1$ for $g \in G$, the module k is called the trivial kG-module. The kernel of the augmentation map is a two-sided ideal $\mathcal{A} = \mathcal{A}_G$ in kG of codimension 1 and has $(g-1)_{g \in G-\{1\}}$ as a k-basis. It is called the augmentation ideal. So we have the exact sequence in ${}_{kG}\mathfrak{mod}$

$$0 \longrightarrow \mathcal{A} \hookrightarrow kG \xrightarrow{\varepsilon} k \longrightarrow 0 .$$

Proposition 1.2 *The augmentation ideal of a finite p-group is nilpotent. (NB. $p = \operatorname{char} k$.)*

Proof. We shall do induction on the order of the p-group G. The statement is trivial if G is trivial, so suppose first that $|G| = p$. Then $G = \langle x \rangle$ and $\left(1, x-1, x^2-1, \ldots, x^{p-1}-1\right)$ is a k-basis for the commutative algebra kG, and $\left(x-1, x^2-1, \ldots, x^{p-1}-1\right)$ is a k-basis for $\mathcal{A}_G$. Hence each element of $\mathcal{A}_G$ is of the form $a(x-1)$ for some $a \in kG$. But $(x-1)^p = x^p - 1 = 1 - 1 = 0$. So $\mathcal{A}_G^p = 0$.

Suppose now that $|G| > p$. Let $H \subsetneq G$ be a maximal subgroup. Then H is a normal subgroup of index p in G. The natural projection $G \xrightarrow{\sigma} G/H$ induces a surjective homomorphism $kG \xrightarrow{\theta} k(G/H)$. Note that the augmentation map $\varepsilon_G : kG \to k$ factors as $\varepsilon_G = \varepsilon_{G/H} \circ \theta$. So $I := \ker \theta \subseteq \ker \varepsilon_G = \mathcal{A}_G$, and moreover $\theta(\mathcal{A}_G) = \theta(\ker \varepsilon_G) = \theta(\ker \varepsilon_{G/H} \circ \theta) = \ker \varepsilon_{G/H} = \mathcal{A}_{G/H}$. We have already proved that $\mathcal{A}_{G/H}^p = 0$. Hence $\mathcal{A}_G^p \subseteq \ker \theta = I$, and it suffices to prove that I is nilpotent.

Fix an element $x \in G - H$. Then since $G = \bigcup_{i=0}^{p-1} x^i H$, each $a \in kG$ is of the form

$$a = \sum_{i=0}^{p-1} \sum_{h \in H} a_{x^i h}\, x^i h = \sum_{i=0}^{p-1} x^i \sum_{h \in H} a_{x^i h}\, h .$$

If moreover $a \in I$, we have

$$0 = \theta(a) = \sum_{i=0}^{p-1} \sigma(x)^i \sum_{h \in H} a_{x^i h}$$

from which we conclude that $\sum_{h \in H} a_{x^i h} = 0$, that is, $\sum_{h \in H} a_{x^i h}\, h \in \mathcal{A}_H$ for $i = 0, \ldots, p-1$. It follows that $I \subseteq kG \cdot \mathcal{A}_H$, and in fact, $I = kG \cdot \mathcal{A}_H$.

The formula $g_1(h_1 - 1) g_2(h_2 - 1) = g_1 g_2 (g_2^{-1} h_1 g_2 - 1)(h_2 - 1)$ shows that $I^2 = kG \cdot \mathcal{A}_H^2$, and, more generally, $I^n = kG \cdot \mathcal{A}_H^n$. Since $\mathcal{A}_H$ is nilpotent by induction, the proof is finished. □

Corollary 1.3 *If G is a finite p-group, then $\mathcal{A}_G^{|G|} = 0$.*

Proof. The sequence $(\dim \mathcal{A}_G^n)_{n=1,2,\ldots}$ is strictly decreasing as long as n is small enough. □

Corollary 1.4 *If G is a finite p-group, then kG is a local ring with (unique) maximal ideal $\mathcal{A}_G$.*

Corollary 1.5 *If G is a finite p-group and M is a projective module in ${}_{kG}\mathfrak{mod}$, then M is a free kG-module.*

This is a consequence of the fact that projective modules are free for any Noetherian local ring (see (5.24) of [CR1]). Now notice that the restriction of any projective module to a subgroup is projective. Hence we have the following corollary.

Corollary 1.6 *If M is a projective module in ${}_{kG}\mathfrak{mod}$, then $p^n \mid \dim M$, where p^n is the order of a Sylow p-subgroup of G.*

Recall that if M, N are in ${}_{kG}\mathfrak{mod}$, then the k-vector space $\mathrm{Hom}(M, N)$ of all k-linear homomorphisms from M to N can be made into a kG-module:

$$(gf)(m) := g \cdot f(g^{-1}m) \qquad \bigl(\text{for } g \in G,\ f \in \mathrm{Hom}(M,N),\ m \in M\bigr).$$

Recall that the radical $\operatorname{rad} A$ of a finite-dimensional algebra A with 1 over a field is the intersection of all maximal left ideals in A. It is then a two-sided nilpotent ideal, and it coincides with the sum of all nilpotent left ideals in A (see (5.15) of [CR1]). The algebra A is semisimple if and only if $\operatorname{rad} A = 0$.

We can now prove Maschke's Theorem

Theorem 1.7 (Maschke) *The group algebra kG is semisimple if and only if the characteristic of k does not divide the order of G.*

Proof. We will actually prove that kG has a nonzero nilpotent left ideal if and only if $\operatorname{char} k \mid |G|$.

First assume that $\operatorname{char} k \mid |G|$. Let $\widetilde{G} := \sum_{g \in G} g \in kG$. Note that if $x \in G$, then $x\widetilde{G} = \widetilde{G}\ (= \widetilde{G}x)$. So if $a \in kG$, then $a\widetilde{G} = \varepsilon(a)\widetilde{G}$. Hence $k\widetilde{G}$ is a one-dimensional (two-sided) ideal in kG. Since $(\widetilde{G})^2 = |G| \cdot \widetilde{G} = 0$, $k\widetilde{G}$ is a nonzero nilpotent ideal in kG.

Suppose now that $\operatorname{char} k \nmid |G|$. (In particular, $\operatorname{char} k = 0$ is allowed here.) We will show that every exact sequence in ${}_{kG}\mathfrak{mod}$ splits. Let

$$0 \longrightarrow A \xrightarrow{\ \alpha\ } B \xrightarrow{\ \beta\ } C \longrightarrow 0$$

be an exact sequence in ${}_{kG}\mathfrak{mod}$. There is a k-homomorphism $\varphi : C \to B$ such that $\beta\varphi = \mathrm{id}_C$. Define $\widehat{\varphi} : C \to B$ by

$$\widehat{\varphi} := \frac{1}{|G|} \sum_{g \in G} g\varphi,$$

and note that $\widehat{\varphi}$ is a kG-homomorphism and that $\beta\widehat{\varphi} = \mathrm{id}_C$. To finish the proof of the theorem, we need Exercise 1.1 which shows that kG has no nonzero nilpotent left ideal. □

EXERCISE 1.1 If A is a finite-dimensional algebra with 1 over a field and $N \subseteq A$ is a nonzero nilpotent left ideal, then the exact sequence

$$0 \longrightarrow N \longrightarrow A \longrightarrow A/N \longrightarrow 0$$

of A-modules does not split.

2 Tensor products, Homs, and duality

Two of the most fundamental and useful operations on the module category are the functors Hom and $\otimes$ from ${}_{kG}\mathfrak{mod} \times {}_{kG}\mathfrak{mod}$ to ${}_{kG}\mathfrak{mod}$. The two are intimately related through duality, and they can be used to develop some interesting piece of information about the group algebra and its modules. In this section we investigate the duality and a couple of its ramifications. The most important application of the section is the proof that group algebras are self-injective algebras, and hence that the subcategories of projective and injective modules coincide.

Recall that if M, N are in ${}_{kG}\mathfrak{mod}$, then the k-vector space $M \otimes N$ can be made into a kG-module by defining the action of G as

$$g(m \otimes n) := gm \otimes gn \qquad (\text{for } g \in G,\ m \in M,\ n \in N).$$

As an aside we should mention that if A is any k-algebra and M, N are A-modules, then $M \otimes N$ is an $A \otimes A$-module. For $A = kG$ we have the comultiplication $\Delta : kG \to kG \otimes kG$. This is the kG-algebra homomorphism given by $\Delta(g) = g \otimes g$ (for $g \in G$), the diagonal map, which is part of the Hopf algebra structure of kG. Thus the kG-action on $M \otimes N$ is the pullback along Δ of the $kG \otimes kG$-action.

For M in ${}_{kG}\mathfrak{mod}$ we define the k-dual $M^* = \mathrm{Hom}(M, k)$ as the kG-module of the k-linear homomorphisms from M to the trivial module k. The G-action reads

$$(gf)(m) = f(g^{-1}m) \qquad (\text{for } g \in G,\ f \in M^*,\ m \in M).$$

We should note here that this is a specialization of a more general technique. That is, if A is any augmented k-algebra, $M^* = \mathrm{Hom}(M,k)$ is a *right* A-module by

$$(fa)(m) := f(am) \qquad (\text{for } a \in A,\ f \in M^*,\ m \in M).$$

For $A = kG$ we make it a left module by our usual tricks.

Proposition 2.1 *Let M, N be in ${}_{kG}\mathfrak{mod}$. Then the kG-modules* $\mathrm{Hom}(M,N)$ *and* $M^* \otimes N$ *are naturally isomorphic.*

Proof. Define $\theta = \theta_{M,N} : M^* \otimes N \to \mathrm{Hom}(M,N)$ by $\theta(f \otimes n)(m) := f(m)n$ (for $f \in M^*$, $n \in N$, $m \in M$). Everybody knows from linear algebra that θ is a k-linear isomorphism. The proof that θ is kG-linear and the naturality are left as exercises. □

EXERCISE 2.1 Verify that θ in the proof of Proposition 2.1 is a kG-homomorphism.

EXERCISE 2.2 Verify that the isomorphism $M^* \otimes N \xrightarrow{\theta_{M,N}} \mathrm{Hom}(M,N)$ in the proof of Proposition 2.1 is natural, that is, if $M \xrightarrow{\alpha} M'$, $N \xrightarrow{\beta} N'$ are homomorphisms in ${}_{kG}\mathfrak{mod}$, then the diagrams

$$\begin{array}{ccc} \mathrm{Hom}(M,N) & \xleftarrow{\theta_{M,N}} & M^* \otimes N \\ {\scriptstyle\alpha^*}\uparrow & & \uparrow{\scriptstyle\alpha^*\otimes 1} \\ \mathrm{Hom}(M',N) & \xleftarrow{\theta_{M',N}} & M'^* \otimes N \end{array} \quad \text{and} \quad \begin{array}{ccc} \mathrm{Hom}(M,N) & \xleftarrow{\theta_{M,N}} & M^* \otimes N \\ {\scriptstyle\beta_*}\downarrow & & \downarrow{\scriptstyle 1\otimes\beta} \\ \mathrm{Hom}(M,N') & \xleftarrow{\theta_{M,N'}} & M^* \otimes N' \end{array}$$

commute.

Consider the case that $N = M$ in Proposition 2.1. Then $\mathrm{End}\, M = \mathrm{Hom}(M,M) \cong M^* \otimes M$ is a ring. It's an easy exercise to show that the "proper" product on $M^* \otimes M$ (corresponding to that of $\mathrm{End}\, M$) is given by the formula

$$(f \otimes m)(f' \otimes m') = f(m') \cdot f' \otimes m.$$

Notice that if (m_i) is any k-basis for M and (m_i^*) is its dual basis, then $\mathsf{r} := \sum_i m_i^* \otimes m_i$ corresponds to the identity homomorphism $\mathrm{id}_M \in \mathrm{End}\, M$. We have the kG-homomorphisms

$$\begin{array}{rcl} I : k & \longrightarrow & M^* \otimes M \\ 1 & \longmapsto & \mathsf{r} \end{array} \qquad \text{and} \qquad \begin{array}{rcl} \mathrm{Tr} : M^* \otimes M & \longrightarrow & k \\ f \otimes m & \longmapsto & f(m)\,. \end{array}$$

Notation If M, N are kG-modules, we will write $M \mid N$ to mean that M is isomorphic to a direct summand of N.

Lemma 2.2 *Let M be in ${}_{kG}\mathfrak{mod}$. If $p \nmid \dim M$, then $k \mid M^* \otimes M$.*

Proof. In fact, $\frac{1}{\dim M} I$ is a section (right inverse) of the homomorphism $M^* \otimes M \xrightarrow{\mathrm{Tr}} k$. □

Proposition 2.3 *Let M be in ${}_{kG}\mathfrak{mod}$. Then $M \mid M \otimes M^* \otimes M$. If $p \mid \dim M$, then $M \oplus M \mid M \otimes M^* \otimes M$.*

Proof. If $p \nmid \dim M$, then $M \cong M \otimes k \mid M \otimes M^* \otimes M$ by Lemma 2.2.

So assume $p \mid \dim M$. Let (m_i) be a k-basis of M and (m_i^*) be its dual basis. Define

$$M \otimes M^* \otimes M \underset{\theta}{\overset{\psi}{\rightleftarrows}} M \oplus M$$

by

$$\psi(m \otimes f \otimes m') := \big(f(m)\, m', f(m')\, m\big) \quad \text{and}$$
$$\theta(m, m') := \sum_i m \otimes m_i^* \otimes m_i + \sum_i m_i \otimes m_i^* \otimes m'.$$

Now ψ is onto, and

$$\begin{aligned}\psi\theta(m, m') &= \Big(\underbrace{\sum_i m_i^*(m)\, m_i}_{=m} + \underbrace{\sum_i m_i^*(m_i)}_{=\dim M} m' \,,\, \underbrace{\sum_i m_i^*(m_i)}_{=\dim M} m + \underbrace{\sum_i m_i^*(m')\, m_i}_{=m'}\Big) \\ &= (m, m').\end{aligned}$$

□

The previous lemma and proposition will prove useful in the next section. The rest of the present section is devoted to showing that kG is self-injective (or quasi-Frobenius). As a consequence, a kG-module is projective if and only if it is injective. We need a definition and proposition which extend a well-known concept from linear algebra.

Definition Let U, V be k-vector spaces. A map $\rho : U \times V \to k$ is a *bilinear pairing* if ρ is linear in both variables. It is *nondegenerate* if $\rho(u, v) = 0$ for all $v \in V$ implies $u = 0$ and $\rho(u, v) = 0$ for all $u \in U$ implies $v = 0$. If U, V are kG-modules, we say ρ is *G-invariant* if $\rho(gu, gv) = \rho(u, v)$ (for $g \in G$, $u \in U$, $v \in V$).

Proposition 2.4 *Suppose that $\rho : U \times V \to k$ is a nondegenerate bilinear pairing of finite-dimensional k-vector spaces U, V. Then $U \cong V^*$. Moreover, if U, V are in ${}_{kG}\mathfrak{mod}$ and ρ is G-invariant, then $U \cong V^*$ as kG-modules.*

Proof. The nondegeneracy of ρ implies that the k-linear map $\sigma : U \to V^*$ defined by $\sigma(u) = \rho(u, \ )$ is an isomorphism. If moreover ρ is G-invariant, then σ is in addition a kG-homomorphism. □

For M in ${}_{kG}\mathfrak{mod}$, the natural k-vector space isomorphism $M \cong M^{**}$ is a kG-homomorphism. This allows us to identify M and M^{**}. Note that even if M and M^* are (nonnaturally) isomorphic as k-vector spaces, they are in general not isomorphic as kG-modules. However, for $M = kG$ they are isomorphic.

Theorem 2.5 *$kG \cong kG^*$ as kG-modules, that is, kG is a Frobenius algebra.*

Proof. The proof is an application of Proposition 2.4. Define a bilinear pairing $kG \times kG \to k$ by $(g, g') \mapsto \delta_{g,g'}$ (Kronecker symbol) for $g, g' \in G$ and extend bilinearly. It is obviously nondegenerate and G-invariant. □

Theorem 2.6 *kG is an injective kG-module, that is, kG is self-injective.*

Proof. Suppose we have a diagram in ${}_{kG}\mathfrak{mod}$

$$\begin{array}{ccccc} 0 & \longrightarrow & A & \xrightarrow{\varphi} & B \\ & & \downarrow{\scriptstyle \sigma} & & \\ & & kG & & \end{array}$$

with exact row. Injectivity of kG means that we can fill in a kG-homomorphism $B \xrightarrow{\psi} kG$ which makes the diagram commutative. To show that we can find such a ψ, we take duals and get the diagram

$$\begin{array}{ccccc} & & kG^* \cong kG & & \\ & {\scriptstyle \theta} \swarrow & \downarrow{\scriptstyle \sigma^*} & & \\ B^* & \xrightarrow{\varphi^*} & A^* & \longrightarrow & 0 \end{array}$$

in ${}_{kG}\mathfrak{mod}$ with exact row. Since $kG^* \cong kG$ by Theorem 2.5, and kG is projective, there is a kG-homomorphism $kG^* \xrightarrow{\theta} B^*$ making the diagram commutative. Now again take duals to get $\psi = \theta^*$. □

Corollary 2.7 *Every (finitely generated) injective kG-module is projective, and every (finitely generated) projective kG-module is injective.*

Exercise 2.3 Prove Corollary 2.7. [Hint: Step 1: The dual of a projective module is injective and vice versa. Step 2: The dual of a projective module is projective.]

3 Restriction and induction

In this section we introduce the restriction and induction functors, which are used extensively in group representations and cohomology. The functors are related to the tensor product operation through the Frobenius Reciprocity Theorem. One consequence of Frobenius reciprocity is the fact that the tensor product of any module with a projective module is again a projective module. This result is extremely important for the homological algebra.

Definition Let M be in ${}_{kG}\mathfrak{mod}$, H a subgroup of G, and L in ${}_{kH}\mathfrak{mod}$. We denote the restriction of M to ${}_{kH}\mathfrak{mod}$ as M_H or possibly $M{\downarrow}_H$. The induced module $L{\uparrow}^G$ in ${}_{kG}\mathfrak{mod}$ is defined as

$$L{\uparrow}^G := kG \otimes_{kH} L$$

with kG acting by left multiplication.

Remarks

(1) As k-vector spaces we have

$$L{\uparrow}^G = \bigoplus_{i=1}^{t} x_i \otimes L \tag{3.1}$$

where $x_1, \ldots, x_t$ is a complete set of representatives of the left cosets of H in G. If $H \trianglelefteq G$, then (3.1) is an equality of kH-modules, since for $h \in H$

$$h(x_i \otimes l) = x_i \otimes (x_i^{-1} h x_i) l.$$

The Mackey formula (see (10.18) of [CR1]) provides a description of $L{\uparrow}^G$ as a kH-module in the case where H is not necessarily normal in G.

(2) The functor

$$\begin{array}{rccc} \mathrm{Ind}_H^G : & {}_{kH}\mathfrak{mod} & \longrightarrow & {}_{kG}\mathfrak{mod} \\ & L & \longmapsto & L{\uparrow}^G \end{array}$$

is exact because kG is a free right kH-module.

Note that

$$\begin{array}{rccc} \mathrm{Res}_H^G : & {}_{kG}\mathfrak{mod} & \longrightarrow & {}_{kH}\mathfrak{mod} \\ & M & \longmapsto & M{\downarrow}_H \end{array}$$

is an exact functor for trivial reasons.

(3) $kG \cong k_{\langle\rangle}{\uparrow}^G$ is the module induced from the trivial subgroup $\langle\rangle$ of the trivial $k\langle\rangle$-module $k_{\langle\rangle}$.

Theorem 3.1 (Frobenius reciprocity) *Let M be in ${}_{kG}\mathfrak{mod}$, H a subgroup of G, and L in ${}_{kH}\mathfrak{mod}$. Then there is a natural isomorphism in ${}_{kG}\mathfrak{mod}$*

$$L{\uparrow}^G \otimes M \cong (L \otimes M{\downarrow}_H){\uparrow}^G.$$

Proof. Define $L{\uparrow}^G \otimes M \overset{\psi}{\underset{\theta}{\rightleftarrows}} (L \otimes M{\downarrow}_H){\uparrow}^G$ by the rules $\psi((g \otimes l) \otimes m) := g \otimes (l \otimes g^{-1}m)$ and $\theta(g \otimes (l \otimes m)) := (g \otimes l) \otimes gm$ (for $g \in G$, $l \in L$, $m \in M$). Since ψ, θ are surely well-defined, inverse to each other, and natural, it remains to be seen that ψ is a kG-homomorphism: for $x, g \in G$, $l \in L$, $m \in M$ we compute

$$\begin{aligned}\psi(x((g \otimes l) \otimes m)) &= \psi((xg \otimes l) \otimes xm) = xg \otimes (l \otimes (xg)^{-1}xm)\\ &= xg \otimes (l \otimes g^{-1}m) = x(g \otimes (l \otimes g^{-1}m)) = x\,\psi((g \otimes l) \otimes m).\end{aligned}$$

□

The Eckmann-Shapiro Lemma, which follows, is very similar in form to Frobenius reciprocity though the proof is quite different.

Proposition 3.2 (Eckmann-Shapiro Lemma) *Let M be in ${}_{kG}\mathfrak{mod}$, H a subgroup of G, and L in ${}_{kH}\mathfrak{mod}$. Then there are natural isomorphisms*

$$\begin{aligned}\mathrm{Hom}_{kG}(L{\uparrow}^G, M) &\cong \mathrm{Hom}_{kH}(L, M{\downarrow}_H),\\ \mathrm{Hom}_{kG}(M, L{\uparrow}^G) &\cong \mathrm{Hom}_{kH}(M{\downarrow}_H, L).\end{aligned}$$

[In other words the functors Ind_H^G and Res_H^G are adjoint functors on both sides.]

Proof. Let $\alpha : L \to L{\uparrow}^G$ and $\beta : L{\uparrow}^G \to L$ be the two kH-homomorphisms defined by $\alpha(l) := 1 \otimes l$ and $\beta(g \otimes l) := \begin{cases} gl & \text{if } g \in H \\ 0 & \text{otherwise} \end{cases}$ (for $g \in G$), respectively.

Define $\mathrm{Hom}_{kG}(L{\uparrow}^G, M) \overset{\psi}{\underset{\theta}{\rightleftarrows}} \mathrm{Hom}_{kH}(L, M{\downarrow}_H)$ by

$$\psi(\sigma) := \sigma\alpha \quad \text{and} \quad \theta(\tau)(g \otimes l) := g\tau(l).$$

Define $\mathrm{Hom}_{kG}(M, L{\uparrow}^G) \overset{\omega}{\underset{\eta}{\rightleftarrows}} \mathrm{Hom}_{kH}(M{\downarrow}_H, L)$ by

$$\omega(\rho) := \beta\rho \quad \text{and} \quad \eta(v)(m) := \sum_{i=1}^{t} x_i \otimes v(x_i^{-1}m)$$

where $x_1, \ldots, x_t$ is a complete set of representatives of the left cosets of H in G. It is an easy exercise to show that these maps are isomorphisms. Check the usual things. □

Consider the set of isomorphism classes of objects in ${}_{kG}\mathfrak{mod}$. The direct sum and the tensor product induce an addition and a multiplication on this set. The next theorem shows that these operations descend if we compute modulo projective kG-modules

Theorem 3.3 *If P is a projective module in ${}_{kG}\mathfrak{mod}$ and M is any kG-module, then $P \otimes M$ is projective.*

Proof. We prove the theorem in the case that M is in ${}_{kG}\mathfrak{mod}$. There is then a kG-module Q such that $P \oplus Q \cong (kG)^n$ is a free kG-module of rank n for some n. We employ Frobenius reciprocity for the trivial subgroup $\langle\rangle \subseteq G$. Let V be a free $k\langle\rangle$-module of rank n, which is nothing but an n-dimensional k-vector space. We have $P \oplus Q \cong (kG)^n \cong V{\uparrow^G}$ and therefore

$$P \otimes M \oplus Q \otimes M \cong (P \oplus Q) \otimes M \cong V{\uparrow^G} \otimes M \cong (V \otimes M_{\langle\rangle}){\uparrow^G} \cong (kG)^{n \cdot \dim M}.$$

So $P \otimes M$ is a direct summand of a free module. □

Notation We will write $L \cong M \oplus (\text{proj})$ to mean that $L \cong M \oplus P$ where P is some projective module.

Lemma 3.4 *Let N be in ${}_{kG}\mathfrak{mod}$. If $N \otimes N \cong N \oplus (\text{proj})$ and $k \,|\, N$, then $N \cong k \oplus (\text{proj})$.*

Proof. Let $N \cong k \oplus L$. From $N \oplus (\text{proj}) \cong N \otimes N \cong (k \oplus L) \otimes (k \oplus L) \cong k \oplus L \oplus L \oplus (L \otimes L) \cong N \oplus L \oplus (L \otimes L)$ we see that L is a direct summand of a projective module. □

Theorem 3.5 *Let M be in ${}_{kG}\mathfrak{mod}$. If $M \otimes M \cong M \oplus (\text{proj})$ and M is nonprojective, then $M \cong k \oplus (\text{proj})$.*

Proof. From $M \otimes M \cong M \oplus (\text{proj})$ we get that $M^* \otimes M^* \cong (M \otimes M)^* \cong \big(M \oplus (\text{proj})\big)^* \cong M^* \oplus (\text{proj})$ and also $(M^* \otimes M) \otimes (M^* \otimes M) \cong M^* \otimes M \oplus (\text{proj})$.

Suppose $p \nmid \dim M$. By Lemma 2.2 we have $k \,|\, M^* \otimes M$, and by taking $N = M^* \otimes M$ in Lemma 3.4 we recognize that $M^* \otimes M \cong k \oplus (\text{proj})$. So

$$M \otimes M^* \otimes M \cong \big(k \oplus (\text{proj})\big) \otimes M \cong M \oplus (\text{proj}). \tag{3.2}$$

On the other hand we have

$$\begin{aligned} M \otimes M^* \otimes M &\cong M^* \otimes M \otimes M \cong M^* \otimes \big(M \oplus (\text{proj})\big) \cong M^* \otimes M \oplus (\text{proj}) \\ &\cong k \oplus (\text{proj}) \end{aligned} \tag{3.3}$$

From (3.2), (3.3) it follows that $M \cong k \oplus (\text{proj})$.

Assume now that $p \mid \dim M$. Then $M \oplus M \mid M \otimes M^* \otimes M$ by Proposition 2.3. On the other hand $M \otimes M^* \otimes M \cong M^* \otimes M \oplus (\text{proj})$ as in the first line of equation (3.3). Hence $M \oplus M \mid M^* \otimes M \oplus (\text{proj})$. Taking n^{th} tensor powers we get

$$\begin{array}{ccc} (M \oplus M)^{\otimes n} & \mid & (M^* \otimes M \oplus (\text{proj}))^{\otimes n} \\ \cong & & \cong \\ M^{\oplus 2^n} \oplus (\text{proj}) \cong (M^{\otimes n})^{\oplus 2^n} & & M^* \otimes M \oplus (\text{proj}) \end{array}$$

Hence $M^* \otimes M$ contains at least 2^n direct summands. But n can be chosen arbitrarily large. Hence the case $p \mid \dim M$ does not occur. □

There are a few other facts about projectives which will be useful. These will not play a big role in our theory but may be needed occasionally. We omit the proofs because they are not particularly important for what we will do later.

We already know that kG is a Frobenius algebra. Since kG is even a symmetric algebra, the head and the socle of each principal indecomposable kG-module are isomorphic.

Theorem 3.6 *Let P be an indecomposable projective kG-module. Then* $\operatorname{soc} P \cong P/\operatorname{rad} P$ *is a simple kG-module. Every simple kG-module is isomorphic to* $\operatorname{soc} P \cong P/\operatorname{rad} P$ *for some indecomposable projective module P.*

We refer to [CR1, §9A] for a proof.

4 Projective resolutions and cohomology

Definition A *projective cover* of a module M is a projective module P_M together with a surjective homomorphism $P_M \overset{\varepsilon}{\twoheadrightarrow} M$ satisfying the following property: if $Q \overset{\theta}{\twoheadrightarrow} M$ is a homomorphism from a projective module Q onto M, then there is an injective homomorphism $P_M \overset{\sigma}{\hookrightarrow} Q$ such that $\varepsilon = \theta\sigma$. In fact, it can be seen from the proof of the next theorem that any homomorphism $\sigma : P_M \to Q$ which satisfies $\varepsilon = \theta\sigma$ must be injective. Likewise any $\tau : Q \to P_M$ which satisfies $\varepsilon\tau = \theta$ must be surjective.

If $P_M \overset{\varepsilon}{\twoheadrightarrow} M$ is a projective cover of M, then no proper projective submodule of P_M is mapped onto M. Note that projective covers, if they exist, are unique up to isomorphism.

Theorem 4.1 *Let M be in ${}_{kG}\mathfrak{mod}$. Then M has a projective cover.*

Proof. The easiest way of proving the theorem would be to invoke the fact that every module has an injective hull. So take an injective hull of M^* and dualize. However, we follow another route. The argument given below will be of use later.

Choose P_M to be a projective module in ${}_{kG}\mathfrak{mod}$ of smallest dimension and such that there exist $P_M \overset{\varepsilon}{\twoheadrightarrow} M$ (such a P_M exists because M is a quotient of a free module of finite rank). Now suppose we are given Q and θ as in the definition above. Since P_M and Q are projective, there is a commutative diagram

$$\begin{array}{ccc} P_M & \overset{\sigma}{\underset{\tau}{\rightleftarrows}} & Q \\ & {\scriptstyle\varepsilon}\searrow \qquad \swarrow{\scriptstyle\theta} & \\ & M & \end{array}$$

Let $\psi := \tau\sigma : P_M \to P_M$. We claim that ψ is an automorphism. Since P_M is finite-dimensional, we have $P_M = \ker \psi^n \oplus \operatorname{im} \psi^n$ for n sufficiently large (Fitting's Lemma). So $\operatorname{im} \psi^n$ is projective. Moreover, $\varepsilon \circ \psi^n = \varepsilon$ by the commutativity of the diagram. By minimality we have $\ker \psi^n = 0$, that is, ψ is an automorphism. Hence σ is injective. □

Proposition 4.2 (Schanuel's Lemma) *Suppose that* $P \overset{\varepsilon}{\twoheadrightarrow} M$ *and* $Q \overset{\theta}{\twoheadrightarrow} M$ *are two homomorphisms onto* M *with* P *and* Q *projective. Then* $\ker\varepsilon \oplus Q \cong \ker\theta \oplus P$.

Proof. We have the commutative diagram with exact rows and columns,

$$\begin{array}{ccccccccc} & & & & 0 & & 0 & & \\ & & & & \downarrow & & \downarrow & & \\ & & & & \ker\theta & = & \ker\theta & & \\ & & & & \downarrow & & \downarrow & & \\ 0 & \longrightarrow & \ker\varepsilon & \longrightarrow & B & \longrightarrow & Q & \longrightarrow & 0 \\ & & \| & & \downarrow & \text{pullback} & \downarrow{\scriptstyle\theta} & & \\ 0 & \longrightarrow & \ker\varepsilon & \longrightarrow & P & \underset{\varepsilon}{\longrightarrow} & M & \longrightarrow & 0 \\ & & & & \downarrow & & \downarrow & & \\ & & & & 0 & & 0 & & \end{array}$$

where $B = \{(p,q) \in P \oplus Q \mid \varepsilon(p) = \theta(q)\}$ is the pullback of the diagram defined by ε and θ. It's an easy exercise to fill in all of the maps. Since P and Q are projective, the middle row and column split, that is, $B \cong \ker\theta \oplus P$ and $B \cong \ker\varepsilon \oplus Q$. □

Definition A *minimal projective resolution* of a module M is a projective resolution

$$\cdots \xrightarrow{\partial_{n+1}} P_n \xrightarrow{\partial_n} P_{n-1} \xrightarrow{\partial_{n-1}} \cdots \to P_0 \xrightarrow{\varepsilon} M \to 0$$

(or in shorthand $P_* \overset{\varepsilon}{\twoheadrightarrow} M$) such that if $Q_* \twoheadrightarrow M$ is another projective resolution of M, then there is an injective chain map $\mu_* : (P_* \overset{\varepsilon}{\twoheadrightarrow} M) \to (Q_* \twoheadrightarrow M)$ and likewise a surjective chain map $\mu'_* : (Q_* \twoheadrightarrow M) \to (P_* \overset{\varepsilon}{\twoheadrightarrow} M)$ such that both μ_* and μ'_* lift the identity on M.

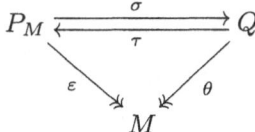

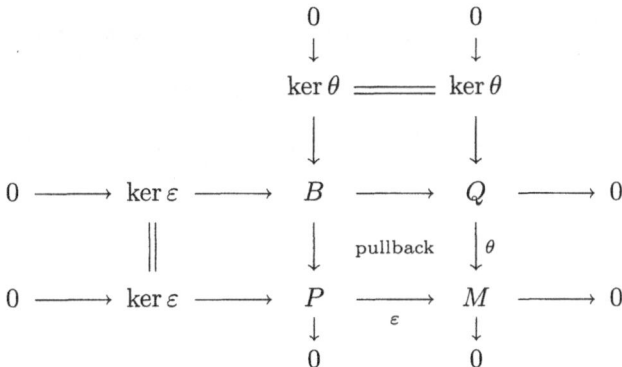

A *minimal injective resolution* of M is an injective resolution

$$0 \longrightarrow M \xrightarrow{\theta} I_0 \longrightarrow \cdots \xrightarrow{\partial^{n-1}} I_{-(n-1)} \xrightarrow{\partial^n} I_{-n} \xrightarrow{\partial^{n+1}} \cdots$$

(or in shorthand $M \overset{\theta}{\hookrightarrow} I_*$) such that if $M \hookrightarrow J_*$ is another injective resolution of M, then there is an injective chain map $\nu_* : (M \overset{\theta}{\hookrightarrow} I_*) \to (M \hookrightarrow J_*)$ and likewise a surjective chain map $\nu'_* : (M \hookrightarrow J_*) \to (M \overset{\theta}{\hookrightarrow} I_*)$ such that both ν_* and ν'_* cover the identity on M.

Theorem 4.1 shows that minimal projective resolutions exist. In fact, let $P_0 \overset{\varepsilon}{\twoheadrightarrow} M$ be a projective cover of M, $P_1 \twoheadrightarrow \ker \varepsilon$ a projective cover of $\ker \varepsilon$, and so on. To obtain a minimal injective resolution we proceed similarly by successively constructing injective hulls.

Theorem 4.3 *Let M be in ${}_{kG}\mathfrak{mod}$. Then M has a minimal projective resolution and a minimal injective resolution.*

Definition Let $P_* \overset{\varepsilon}{\twoheadrightarrow} M$ be a minimal projective resolution and $M \overset{\theta}{\hookrightarrow} I_*$ a minimal injective resolution of M. We define for $n > 0$

$$\Omega^n(M) := \ker \partial_{n-1} = \operatorname{im} \partial_n \cong \operatorname{coim} \partial_n = \operatorname{coker} \partial_{n+1}$$

(where $\ker \partial_0 := \ker \varepsilon$), and for $n = 1$ we abbreviate $\Omega^1(M)$ by $\Omega(M)$. Further, let

$$\Omega^{-n}(M) := \operatorname{coker} \partial^{n-1} = \operatorname{coim} \partial^n \cong \operatorname{im} \partial^n = \ker \partial^{n+1}$$

(where $\operatorname{coker} \partial^0 := \operatorname{coker} \theta$). For $n = 0$ we let

$$\Omega^0(M) := \Omega^{-1}(\Omega(M)),$$

so that $M \cong \Omega^0(M) \oplus (\text{proj})$.

Schanuel's Lemma shows that the modules $\Omega^n(M)$ are well-defined up to isomorphism. If we drop the adjective "minimal" in the definition of the modules $\Omega^n(M)$, we get modules $\widetilde{\Omega}^n(M) \cong \Omega^n(M) \oplus (\text{proj})$.

Suppose that $P_* \overset{\varepsilon}{\twoheadrightarrow} M$ and $M \overset{\theta}{\hookrightarrow} I_*$ are a minimal projective and a minimal injective resolution of M, respectively. We can splice them together and get in this way a minimal complete resolution of M. It looks like this:

$$\begin{array}{ccccccccccccc}
\cdots \longrightarrow & P_1 & & \xrightarrow{\partial_1} & & P_0 & \xrightarrow{\partial_0} & P_{-1} & & \xrightarrow{\partial_{-1}} & & P_{-2} & \longrightarrow \cdots \\
& & \searrow & & \nearrow & & {\scriptstyle\varepsilon}\searrow \quad \nearrow{\scriptstyle\theta} & & \searrow & & \nearrow & & \\
& & & \Omega^1(M) & & & M & & & \Omega^{-1}(M) & & &
\end{array}$$

where $P_{-n} := I_{-(n-1)}$ and $\partial_{-n} := \partial^n$ for $n > 0$, and $\partial_0 := \theta\varepsilon$.

Proposition 4.4 *Let M, N be in ${}_{kG}\mathfrak{mod}$, H a subgroup of G, L in ${}_{kH}\mathfrak{mod}$, and n, m integers. Then*

(i) $\Omega^n(\mathrm{proj}) = 0$.

(ii) $\Omega^n(M)$ *has no nonzero projective submodules.*

(iii) $\Omega^n(M \oplus N) \cong \Omega^n(M) \oplus \Omega^n(N)$.

(iv) $\Omega^n(M)^* \cong \Omega^{-n}(M^*)$.

(v) $\Omega^n\big(\Omega^m(M)\big) \cong \Omega^{n+m}(M)$.

(vi) $\Omega^m(M) \otimes \Omega^n(N) \cong \Omega^{n+m}(M \otimes N) \oplus (\mathrm{proj})$.

(vii) $\Omega^n(M){\downarrow_H} \cong \Omega^n(M{\downarrow_H}) \oplus (\mathrm{proj})$.

(viii) $\Omega^n(L){\uparrow^G} \cong \Omega^n\big(L{\uparrow^G}\big) \oplus (\mathrm{proj})$.

Proof. In view of (iv) (for $n > 0$) and the definition of Ω^0 we may assume that n, m are positive integers.

(i) and (ii) are trivial. (iii) follows from (ii) together with Schanuel's Lemma. (iv) is clear by duality. (v) follows by truncating a minimal projective resolution of M to a minimal projective resolution of $\Omega^m(M)$.

For (vi) let $\cdots \to P_n \xrightarrow{\partial_n} P_{n-1} \to \cdots \to P_0 \xrightarrow{\varepsilon} M \to 0$ be a minimal projective resolution of M. By tensoring this exact sequence with N we get the exact sequence $\cdots \to P_n \otimes N \xrightarrow{\partial_n \otimes 1} P_{n-1} \otimes N \to \cdots \to P_0 \otimes N \xrightarrow{\varepsilon \otimes 1} M \otimes N \to 0$, which is a (not necessarily minimal) projective resolution of $M \otimes N$. So

$$\Omega^m(M) \otimes N \cong \mathrm{im}(\partial_m \otimes 1) \cong \Omega^m(M \otimes N) \oplus (\mathrm{proj}). \tag{4.1}$$

Similarly we have that

$$\Omega^m(M) \otimes \Omega^n(N) \cong \Omega^n\big(\Omega^m(M) \otimes N\big) \oplus (\mathrm{proj}). \tag{4.2}$$

Putting (4.1) and (4.2) together we arrive at

$$\begin{aligned}\Omega^m(M) \otimes \Omega^n(N) &\cong \Omega^n\big(\Omega^m(M \otimes N) \oplus (\mathrm{proj})\big) \oplus (\mathrm{proj}) \\ &\cong \Omega^n\big(\Omega^m(M \otimes N)\big) \oplus (\mathrm{proj}) \cong \Omega^{n+m}(M \otimes N) \oplus (\mathrm{proj}).\end{aligned}$$

Since the restriction of a minimal projective resolution of M is a (not necessarily minimal) projective resolution of $M{\downarrow_H}$, we get (vii).

If $Q_* \overset{\theta}{\twoheadrightarrow} L$ is a minimal projective resolution of L, then $kG \otimes_{kH} Q_* \overset{1\otimes\theta}{\twoheadrightarrow} L{\uparrow^G}$ is a (not necessarily minimal) projective resolution of $L{\uparrow^G}$. So (viii) follows. □

Definition Let M, N be in ${}_{kG}\mathfrak{mod}$. Let $P_* \twoheadrightarrow M$ be a projective resolution of M. Applying $\mathrm{Hom}_{kG}(\ \ , N)$ we get the complex

$$0 \longrightarrow \mathrm{Hom}_{kG}(P_0, N) \longrightarrow \mathrm{Hom}_{kG}(P_1, N) \longrightarrow \ \cdots\ .$$

Then the Ext^n_{kG} functor is defined as the cohomology of the complex:

$$\mathrm{Ext}^n_{kG}(M, N) := H^n\big(\mathrm{Hom}_{kG}(P_*, N)\big).$$

If $M = k$ is the trivial kG-module, then we have a special notation:

$$H^n(G, N) := \mathrm{Ext}^n_{kG}(k, N).$$

This is called the *cohomology of G with coefficients in N.*

We will assume that the reader has some familiarity with cohomology. In particular, the Ext functor does not depend on the choice of the projective resolution. Moreover

$$\mathrm{Ext}^n_{kG}(M, N) \cong H^n\big(\mathrm{Hom}_{kG}(M, I_*)\big)$$

where $N \hookrightarrow I_*$ is an injective resolution of N. See [HS] for details.

Theorem 4.5 *Let M, N be in ${}_{kG}\mathfrak{mod}$ and let n be a positive integer.*

(i) *Every cohomology element $\zeta \in \mathrm{Ext}^n_{kG}(M, N)$ is represented by a homomorphism $\bar{\zeta} : \Omega^n(M) \to N$.*

(ii) *Every homomorphism $\hat{\zeta} : \Omega^n(M) \to N$ represents a cohomology element* $\mathrm{class}(\hat{\zeta}) \in \mathrm{Ext}^n_{kG}(M, N)$.

(iii) *Two such homomorphisms $\hat{\zeta}, \tilde{\zeta}$ represent the same class if and only if $\hat{\zeta} - \tilde{\zeta}$ factors through a projective kG-module.*

Proof. (i) Let $P_* \stackrel{\varepsilon}{\twoheadrightarrow} M$ be a minimal projective resolution of M. Now ζ is represented by a cocycle $\zeta' : P_n \to N$. Being a cocycle means that $\zeta' \circ \partial_{n+1} = 0$. Hence ζ' factors through $\mathrm{coker}\, \partial_{n+1} \cong \Omega^n(M)$.

$$\begin{array}{ccccccc}
\cdots & \longrightarrow & P_{n+1} & \xrightarrow{\partial_{n+1}} & P_n & \longrightarrow & \cdots \\
 & & & & {\scriptstyle\zeta'}\big\downarrow \ \searrow^{q} & & \\
 & & & & \quad\ \ \Omega^n(M) & & \\
 & & & & \ \ \swarrow_{\bar{\zeta}} & & \\
 & & & & N & &
\end{array}$$

(ii) If we are given $\hat{\zeta} : \Omega^n(M) \to N$, then $\hat{\zeta} \circ q = \hat{\zeta} \circ \partial_n : P_n \to N$ represents a cohomology element $\mathrm{class}(\hat{\zeta}) := \mathrm{class}(\hat{\zeta} \circ q) \in \mathrm{Ext}^n_{kG}(M, N)$.

(iii) If $\mathrm{class}(\hat{\zeta}) = \mathrm{class}(\tilde{\zeta})$, then $(\hat{\zeta} - \tilde{\zeta}) \circ q = \eta \circ \partial_n$ for some $\eta : P_{n-1} \to N$. So $\hat{\zeta} - \tilde{\zeta} = \eta\iota$ factors through P_{n-1}, where $\Omega^n(M) \overset{\iota}{\hookrightarrow} P_{n-1}$ is the inclusion.

Suppose that $\varphi := \hat{\zeta} - \tilde{\zeta} : \Omega^n(M) \to N$ factors through a projective module P, say $\varphi = \beta\alpha$. We shall show that φ is a coboundary, i. e., that it factors through P_{n-1}. Consider the diagram

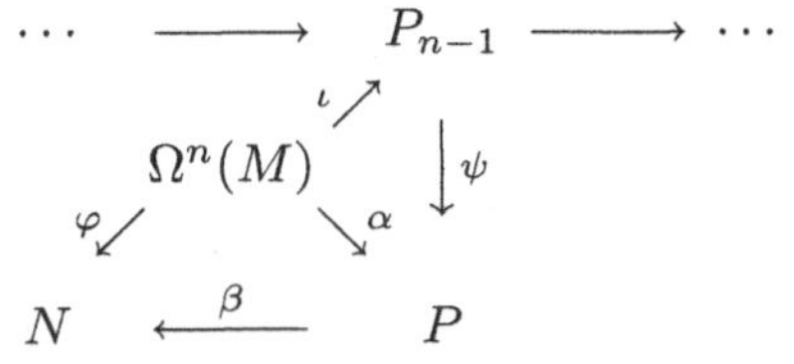

Since P is injective, we have a homomorphism $\psi : P_{n-1} \to P$ with $\psi\iota = \alpha$. So $\varphi = \beta\psi\iota$ factors through P_{n-1}. □

One of the things that we have proved is half of the following proposition.

Proposition 4.6 *If $M \xrightarrow{\alpha} N$ factors through a projective module, then it factors through any injection of M into a projective module and also through any surjection of a projective module onto N.*

This leads naturally to the next section.

5 The stable category

There are several natural homes for doing cohomology theory. For instance, any of several derived categories can be used for this purpose. Here we will focus on the stable category, which has the advantage of being very closely related to the module category.

Definition Let $_{kG}\mathfrak{stmod}$ denote the category of finitely generated left kG-modules modulo projectives, that is, the *stable category.* The objects of $_{kG}\mathfrak{stmod}$ are the same as those of $_{kG}\mathfrak{mod}$. If M, N are in $_{kG}\mathfrak{mod}$, let $\mathrm{PHom}_{kG}(M, N)$ be the subspace of $\mathrm{Hom}_{kG}(M, N)$ consisting of all those kG-homomorphisms which factor through projective modules. Now we define

$$\mathrm{Hom}_{_{kG}\mathfrak{stmod}}(M, N) := \underline{\mathrm{Hom}}_{kG}(M, N) := \mathrm{Hom}_{kG}(M, N) / \mathrm{PHom}_{kG}(M, N).$$

It is an easy exercise to check that $_{kG}\mathfrak{stmod}$ is a category. Of course we have a natural functor $_{kG}\mathfrak{mod} \to {}_{kG}\mathfrak{stmod}$ which is the identity on objects and projects morphisms to the cosets modulo the corresponding PHom_{kG}-subspaces. Theorem 4.5 can now be expressed as follows.

Theorem 5.1 *Let M, N be in ${}_{kG}\mathfrak{mod}$. Then $\operatorname{Ext}^n_{kG}(M,N) \cong \underline{\operatorname{Hom}}_{kG}(\Omega^n(M), N)$ for any positive integer n.*

Definition Let P_*, Q_* be nonnegative complexes, and suppose that we are given two chain maps $\mu_*, \nu_* : P_* \rightrightarrows Q_*$. We say μ_*, ν_* are *homotopic in positive degrees* if there exist maps $s_i : P_i \to Q_{i+1}$ such that

$$\mu_i - \nu_i = s_{i-1} \circ \partial_i^P + \partial_{i+1}^Q \circ s_i \qquad \text{(for all } i > 0\text{)}.$$

We denote by $\mathcal{C}(P_*, Q_*)$ the classes of chain maps $\nu_* : P_* \to Q_*$ under the relation of homotopy in positive degrees.

Proposition 5.2 *Let $P_* \xrightarrow{\varepsilon} M$, $Q_* \xrightarrow{\theta} N$ be projective resolutions of the modules M, N in ${}_{kG}\mathfrak{mod}$, respectively. Then for every integer n there are natural isomorphisms*

$$\underline{\operatorname{Hom}}_{kG}(M,N) \cong \mathcal{C}(P_*, Q_*) \cong \underline{\operatorname{Hom}}_{kG}(\Omega^n(M), \Omega^n(N)).$$

Proof. We indicate how to construct the maps. Without loss of generality we may assume that $n > 1$.

Let $\zeta \in \underline{\operatorname{Hom}}_{kG}(M,N)$. Then any representative $\alpha \in \operatorname{Hom}_{kG}(M,N)$ of ζ can be lifted to a chain map $\mu_* : P_* \to Q_*$. Suppose that $\beta \in \operatorname{Hom}_{kG}(M,N)$ also represents ζ, i.e., that $\alpha - \beta$ factors through a projective. Let $\nu_* : P_* \to Q_*$ be a lift of β. We have the commutative diagram (without the maps s_n)

$$\begin{array}{ccccccccc}
\cdots & \longrightarrow & P_1 & \xrightarrow{\partial_1^P} & P_0 & \xrightarrow{\varepsilon} & M & \longrightarrow & 0 \\
 & & \big\downarrow{\scriptstyle \mu_1-\nu_1} & \swarrow{\scriptstyle s_0} & \big\downarrow{\scriptstyle \mu_0-\nu_0} & \swarrow{\scriptstyle s_{-1}} & \big\downarrow{\scriptstyle \alpha-\beta} & & \\
\cdots & \longrightarrow & Q_1 & \xrightarrow[\partial_1^Q]{} & Q_0 & \xrightarrow[\theta]{} & N & \longrightarrow & 0
\end{array}$$

Since $\alpha - \beta$ factors through a projective, it factors through Q_0 (see Proposition 4.6), say $\theta \circ s_{-1} = \alpha - \beta$. The commutativity of the diagram shows that we have $\theta \circ (\mu_0 - \nu_0 - s_{-1} \circ \varepsilon) = 0$, that is, $\operatorname{im}(\mu_0 - \nu_0 - s_{-1} \circ \varepsilon) \subseteq \ker \theta = \operatorname{im} \partial_1^Q$. Since P_0 is projective, $\mu_0 - \nu_0 - s_{-1} \circ \varepsilon$ factors through Q_1, say

$$\mu_0 - \nu_0 = s_{-1} \circ \varepsilon + \partial_1^Q \circ s_0.$$

Continuing this way—using the projectivity of P_n and the exactness of the lower row at Q_n—we get $s_n : P_n \to Q_{n+1}$ with

$$\mu_n - \nu_n = s_{n-1} \circ \partial_n^P + \partial_{n+1}^Q \circ s_n \qquad (n > 0).$$

Hence we have constructed a map $\underline{\operatorname{Hom}}_{kG}(M,N) \to \mathcal{C}(P_*, Q_*)$.

Let $\eta \in \mathcal{C}(P_*, Q_*)$. Then any chain map $\mu_* : P_* \to Q_*$ representing η restricts to the maps

$$\Omega^n(M) \oplus (\text{proj}) \cong \ker \partial^P_{n-1} \xrightarrow{\mu'_{n-1}} \ker \partial^Q_{n-1} \cong \Omega^n(N) \oplus (\text{proj}).$$

Suppose that $\nu_* : P_* \to Q_*$ is another representative of η. So μ_* and ν_* are chain homotopic in positive degrees. That is, $\mu_{n-1} - \nu_{n-1} = s_{n-2} \circ \partial + \partial \circ s_{n-1}$ for $n > 1$. Hence $\mu'_{n-1} - \nu'_{n-1} = \partial \circ s_{n-1}$ factors through the projective module Q_n. This gives us a map $\mathcal{C}(P_*, Q_*) \to \underline{\mathrm{Hom}}_{kG}(\Omega^n(M), \Omega^n(N))$ for $n > 1$.

Similarly we can map $\underline{\mathrm{Hom}}_{kG}(\Omega^n(M), \Omega^n(N))$ to the set of homotopy classes in negative degrees from any injective resolution of $\Omega^n(M)$ to any injective resolution of $\Omega^n(N)$. In turn this is mapped to $\underline{\mathrm{Hom}}_{kG}(\Omega^{-m}(\Omega^n(M)), \Omega^{-m}(\Omega^n(N)))$ for $m > 1$. In particular, for $m = n$ we have $\Omega^{-n}(\Omega^n(M)) \cong M$, $\Omega^{-n}(\Omega^n(N)) \cong N$ in ${}_{kG}\mathfrak{stmod}$. So we have got a map $\underline{\mathrm{Hom}}_{kG}(\Omega^n(M), \Omega^n(N)) \to \underline{\mathrm{Hom}}_{kG}(M, N)$. □

Proposition 5.3 *Let $M \xrightarrow{\alpha} N$ be a morphism in ${}_{kG}\mathfrak{mod}$. Then there exist projective modules P, Q and modules L, L' in ${}_{kG}\mathfrak{mod}$ such that there are exact sequences*

$$0 \to M \xrightarrow{\alpha'} N \oplus Q \xrightarrow{\gamma} L' \to 0 \qquad \text{and} \qquad 0 \to L \xrightarrow{\beta} M \oplus P \xrightarrow{\alpha''} N \to 0$$

with $\mathrm{pr}_N \circ \alpha' \equiv \alpha''|_M \equiv \alpha \mod \mathrm{PHom}_{kG}(M, N)$ for pr_N being the projection onto N. Moreover we can assume that $L \cong \Omega(L') \oplus (\text{proj})$.

Proof. Let $\ker \alpha \hookrightarrow Q$ be an injective hull of $\ker \alpha$. So we have a commutative diagram

$$\begin{array}{ccc} \ker \alpha & \longrightarrow & M \\ \big\downarrow & \swarrow_{\theta} & \\ Q & & \end{array}$$

Since $\alpha' := \binom{\alpha}{\theta} : M \to N \oplus Q$ is an injective map, we get the exact sequence

$$0 \to M \xrightarrow{\alpha'} N \oplus Q \xrightarrow{\gamma} L' \to 0$$

with $L' := \operatorname{coker} \alpha'$.

Let $P' \twoheadrightarrow L'$ be a projective cover of L'. Then we have the commutative diagram with exact rows

$$\begin{array}{ccccccccc} 0 & \longrightarrow & \Omega(L') & \xrightarrow{\mu} & P' & \longrightarrow & L' & \longrightarrow & 0 \\ & & \big\downarrow{\scriptstyle\varphi} & & \big\downarrow{\scriptstyle\psi} & & \Big\| & & \\ 0 & \longrightarrow & M & \xrightarrow{\alpha'} & N \oplus Q & \longrightarrow & L' & \longrightarrow & 0 \end{array}$$

where ψ exists because of the projectivity of P', and φ is induced by ψ. So we have a sequence

$$0 \to \Omega(L') \xrightarrow{\binom{\varphi}{\mu}} M \oplus P' \xrightarrow{(\alpha' \ -\psi)} N \oplus Q \to 0 \tag{5.1}$$

which can be checked to be exact. Now we can split off the projective module Q. But this means that we have to allow to split off a projective summand of M. In order to restore M, we add this projective summand to $\Omega(L')$. □

EXERCISE 5.1 Show that the sequence (5.1) is exact.

Lemma 5.4 *Let C_*, D_* be nonnegative chain complexes and $\mu_* : C_* \to D_*$ be a chain map. Then there exists a totally split exact nonnegative chain complex D'_* of projective modules and a chain map $\mu'_* : C_* \to D'_*$ such that*

$$\binom{\mu_*}{\mu'_*} : C_* \to D_* \oplus D'_*$$

is injective.

Proof. Let n be the least index such that μ_n in not injective. There exists a projective module P_n and a morphism $C_n \xrightarrow{\mu'_n} P_n$ such that $\binom{\mu_n}{\mu'_n} : C_n \to D_n \oplus P_n$ is injective. (Choose P_n to be an injective hull of $\ker \mu_n$ [cf. the beginning of the proof of Proposition 5.3].) Therefore we have a commutative diagram

$$\begin{array}{ccccccccc}
\cdots \longrightarrow & C_{n+2} & \xrightarrow{\partial} & C_{n+1} & \xrightarrow{\partial} & C_n & \xrightarrow{\partial} & C_{n-1} & \longrightarrow \cdots \\
& \downarrow \mu_{n+2} & & \downarrow \binom{\mu_{n+1}}{\mu'_n \circ \partial} & & \downarrow \binom{\mu_n}{\mu'_n} & & \downarrow \mu_{n-1} & \\
\cdots \longrightarrow & D_{n+2} & \xrightarrow[\binom{\partial}{0}]{} & D_{n+1} \oplus P_n & \xrightarrow[\begin{pmatrix}\partial & 0\\ 0 & 1\end{pmatrix}]{} & D_n \oplus P_n & \xrightarrow[(\partial\ 0)]{} & D_{n-1} & \longrightarrow \cdots
\end{array}$$

If the maps $\ldots, \mu_{n+3}, \mu_{n+2}, \binom{\mu_{n+1}}{\mu'_n \circ \partial}$ are all injective, then we are done. Otherwise repeat the process. □

Remark There are numerous variations of Lemma 5.4. For example, with the same hypothesis there is a totally split exact complex C'_* of projective modules and a chain map $\mu'_* : C'_* \to D_*$ such that

$$(\mu_* \ \mu'_*) : C_* \oplus C'_* \to D_*$$

is surjective. It may be necessary here to allow C'_{-1} to be nonzero.

EXERCISE 5.2 Prove the above remark.

Proposition 5.5 *In Proposition 5.3 the isomorphism classes of β, γ, and L (and also L') in ${}_{kG}\mathfrak{stmod}$ are completely determined by the class of α in ${}_{kG}\mathfrak{stmod}$.*

Proof. Let $M \xrightarrow{\alpha} N$ be a morphism in ${}_{kG}\mathfrak{mod}$. Suppose that

$$0 \to M \xrightarrow{\hat\alpha} N \oplus \hat Q \to L' \to 0 \qquad \text{and} \qquad 0 \to M \xrightarrow{\tilde\alpha} N \oplus \tilde Q \to U \to 0$$

are exact sequences with $\mathrm{pr}_N \circ \hat\alpha \equiv \mathrm{pr}_N \circ \tilde\alpha \equiv \alpha \mod \mathrm{PHom}_{kG}(M, N)$ and where $\hat Q, \tilde Q$ are projective modules. By letting $N \oplus \hat Q$ take the role of N, we may assume that α is injective. More precisely, we are given a diagram with exact rows

$$\begin{array}{ccccccccc} 0 & \longrightarrow & M & \xrightarrow{\alpha} & N & \xrightarrow{\gamma} & L' & \longrightarrow & 0 \\ & & \downarrow{\scriptstyle \rho_M} & & \downarrow{\scriptstyle \rho_N} & & & & \\ 0 & \longrightarrow & M \oplus \hat{Q} & \xrightarrow{\alpha'} & N \oplus \hat{P} & \xrightarrow{\gamma'} & U & \longrightarrow & 0 \end{array} \tag{5.2}$$

where ρ_M, ρ_N are the natural inclusions and where $\hat{Q}, \hat{P}$ are projective modules. Moreover $\rho_N \circ \alpha \equiv \alpha' \circ \rho_M \mod \mathrm{PHom}_{kG}(M, N \oplus \hat{P})$.

We must show that $L' \cong U$ in ${}_{kG}\mathbf{stmod}$ and that the classes of γ and γ' coincide. According to Proposition 5.2 it suffices to show these equivalences after applying the functor Ω^n for some integer n.

Let $P'_* \twoheadrightarrow M$, $P_* \twoheadrightarrow N$, $P''_* \twoheadrightarrow L'$, $Q'_* \twoheadrightarrow M \oplus \hat{Q}$, $Q_* \twoheadrightarrow N \oplus \hat{P}$, and $Q''_* \twoheadrightarrow U$ be projective resolutions such that we have commutative diagrams with exact rows

$$\begin{array}{ccccccccc} 0 & \longrightarrow & P'_* & \xrightarrow{\alpha_*} & P_* & \longrightarrow & P''_* & \longrightarrow & 0 \\ & & \downarrow & & \downarrow & & \downarrow & & \\ 0 & \longrightarrow & M & \xrightarrow{\alpha} & N & \longrightarrow & L' & \longrightarrow & 0 \end{array}$$

and

$$\begin{array}{ccccccccc} 0 & \longrightarrow & Q'_* & \xrightarrow{\alpha'_*} & Q_* & \longrightarrow & Q''_* & \longrightarrow & 0 \\ & & \downarrow & & \downarrow & & \downarrow & & \\ 0 & \longrightarrow & M \oplus \hat{Q} & \xrightarrow{\alpha'} & N \oplus \hat{P} & \longrightarrow & U & \longrightarrow & 0. \end{array}$$

We may further assume that the resolutions $P'_* \twoheadrightarrow M$ and $P''_* \twoheadrightarrow L'$ are minimal.

The square in diagram (5.2) lifts to a diagram of projective resolutions

$$\begin{array}{ccc} P'_* & \xrightarrow{\alpha_*} & P_* \\ \downarrow{\scriptstyle \mu_*} & & \downarrow{\scriptstyle \nu_*} \\ Q'_* & \xrightarrow{\alpha'_*} & Q_* \end{array} \tag{5.3}$$

where $\nu_* \circ \alpha_*$ and $\alpha'_* \circ \mu_*$ are chain homotopic in positive degrees. So we have a diagram

$$\begin{array}{ccccccccc} \cdots & \longrightarrow & P'_{n+1} & \xrightarrow{\partial} & P'_n & \xrightarrow{\partial} & P'_{n-1} & \longrightarrow & \cdots \\ & & \downarrow{\scriptstyle \alpha_{n+1}} & & \downarrow{\scriptstyle \alpha_n} & & \downarrow{\scriptstyle \alpha_{n-1}} & & \\ \cdots & \longrightarrow & P_{n+1} & \xrightarrow{\partial} & P_n & \xrightarrow{\partial} & P_{n-1} & \longrightarrow & \cdots \\ & & \downarrow{\scriptstyle \nu_{n+1}} & & \downarrow{\scriptstyle \nu_n} & & \downarrow{\scriptstyle \nu_{n-1}} & & \\ \cdots & \longrightarrow & Q_{n+1} & \xrightarrow{\partial} & Q_n & \xrightarrow{\partial} & Q_{n-1} & \longrightarrow & \cdots \end{array}$$

and a chain homotopy $P'_n \xrightarrow{s_n} Q_{n+1}$ in positive degrees, so that $\nu_n \circ \alpha_n - \alpha'_n \circ \mu_n = s_{n-1} \circ \partial + \partial \circ s_n$ for $n > 0$. Now each α_n is injective, and Q_{n+1} is an injective module. Hence there are maps $t_n : P_n \to Q_{n+1}$ such that $t_n \circ \alpha_n = s_n$. Thus

$$\begin{aligned}\nu_n \circ \alpha_n - \alpha'_n \circ \mu_n &= s_{n-1} \circ \partial + \partial \circ s_n = t_{n-1} \circ \alpha_{n-1} \circ \partial + \partial \circ t_n \circ \alpha_n \\ &= (t_{n-1} \circ \partial + \partial \circ t_n) \circ \alpha_n\end{aligned}$$

or

$$(\nu_n - t_{n-1} \circ \partial - \partial \circ t_n) \circ \alpha_n = \alpha'_n \circ \mu_n.$$

Note that $\nu_n - t_{n-1} \circ \partial - \partial \circ t_n$ is still a chain map which lifts ρ_N. So replace ν_* by $\nu_* - t_* \circ \partial - \partial \circ t_*$. Now we have $\nu_* \circ \alpha_* = \alpha'_* \circ \mu_*$ in positive degrees, that is, (5.3) commutes in positive degrees. Taking kernels at the $(n-1)^{\text{st}}$ stage, we get

$$\begin{array}{ccccccccc} 0 \to & \Omega^n(M) & \xrightarrow{\Omega^n(\alpha)} & \Omega^n(N) \oplus (\text{proj}) & \xrightarrow{\Omega^n(\gamma)} & \Omega^n(L') & \to 0 \\ & \downarrow{\scriptstyle \mu'_{n-1}} & & \downarrow{\scriptstyle \nu'_{n-1}} & & \downarrow{\scriptstyle \psi_{n-1}} & \\ 0 \to & \Omega^n(M) \oplus (\text{proj}) & \xrightarrow{\Omega^n(\alpha')} & \Omega^n(N) \oplus (\text{proj}) & \xrightarrow{\Omega^n(\gamma')} & \Omega^n(U) \oplus (\text{proj}) & \to 0 \end{array}$$

where ψ_{n-1} is induced from ν'_{n-1}. By adding a suitable exact sequence of projectives to the bottom row, we can make the vertical maps injective. So we get an exact sequence of cokernels whose first two terms are projective, and therefore so is the third. Hence $\Omega^n(U) \oplus (\text{proj}) \cong \Omega^n(L') \oplus (\text{proj})$ and $\Omega^n(\gamma), \Omega^n(\gamma')$ are congruent modulo maps which factor through projectives. This proves one half of the proposition. The other half follows by applying the first half to the dual of the sequence $0 \to L \xrightarrow{\beta} M \oplus P \xrightarrow{\alpha''} N \to 0$. □

Whereas ${}_{kG}\mathfrak{mod}$ is an abelian category, ${}_{kG}\mathfrak{stmod}$ is only a triangulated category. That is, in general there are no kernels and cokernels in ${}_{kG}\mathfrak{stmod}$. In place of this we have the fact that for each morphism in ${}_{kG}\mathfrak{stmod}$ there is (up to isomorphism) a uniquely defined object which is the third object in the triangle of the morphism. This is a key point in the definition of quotient categories, and we may see some of this later on.

Our short term interest in triangulated categories will be more philosophical. The stable category is a natural setting for cohomology, and we want to be familiar with some of the properties of the category. In particular, we will often want to shift from one form of a triangle to another. For more details on triangulated categories the reader is referred to the books by Happel [H] and Weibel [W].

Definition A *triangulated category* is an additive category $\mathfrak{C}$, together with an automorphism $\mathcal{T}$, called the *translation functor*, and a collection of *triangles* which satisfy the following conditions and axioms. Each triangle is a sextuple $(U, V, W, \alpha, \beta, \gamma)$ consisting of objects U, V, and W and morphisms $\alpha : U \to V$, $\beta : V \to W$, and $\gamma : W \to \mathcal{T}U$.

Axiom 1 Any sextuple isomorphic to a triangle is a triangle. Any morphism $\alpha : U \to V$ can be embedded in a unique triangle, $(U, V, W, \alpha, \beta, \gamma)$. The sextuple $(U, U, 0, \mathrm{id}_U, 0, 0)$ is a triangle.

Axiom 2 If $(U, V, W, \alpha, \beta, \gamma)$ is a triangle, then so are $(V, W, \mathcal{T}U, \beta, \gamma, -\mathcal{T}\alpha)$ and $(\mathcal{T}^{-1}W, U, V, -\mathcal{T}^{-1}\gamma, \alpha, \beta)$.

Axiom 3 Given triangles $(U, V, W, \alpha, \beta, \gamma)$ and $(U', V', W', \alpha', \beta', \gamma')$ and morphisms $f : U \to U'$ and $g : V \to V'$ such that $\alpha' f = g\alpha$, then there exists a morphism $h : W \to W'$ such that $\beta' g = h\beta$ and $\gamma' h = \mathcal{T}f \circ \gamma$. (The triple (f, g, h) is a morphism of the triangles.)

Axiom 4 (Octahedral Axiom) Given triangles $(U, V, W, \alpha, \beta, \gamma)$, $(V, W', U', \mu, \nu, \theta)$, and $(U, W', V', \mu\alpha, \tau, \rho)$ then there is a triangle $(W, V', U', f, g, \theta \circ \mathcal{T}\beta)$ such that $g\tau = \nu$, $\rho f = \gamma$, $f\beta = \tau\mu$, and $\mathcal{T}\alpha \circ \rho = \theta g$.

Theorem 5.6 *The stable category* ${}_{kG}\mathfrak{stmod}$ *is a triangulated category with translation functor* $\mathcal{T} = \Omega^{-1}$; *the sextuple* $(U, V, W, \alpha, \beta, \gamma)$ *is a triangle if and only if there exist exact sequences*

$$0 \longrightarrow U \xrightarrow{\alpha'} V \oplus (\mathrm{proj}) \xrightarrow{\beta'} W \longrightarrow 0$$

and

$$0 \longrightarrow V \xrightarrow{\beta''} W \oplus (\mathrm{proj}) \xrightarrow{\gamma'} \Omega^{-1}(U) \longrightarrow 0$$

in ${}_{kG}\mathfrak{mod}$ *such that* $\mathrm{class}(\alpha') = \alpha$, $\mathrm{class}(\beta') = \mathrm{class}(\beta'') = \beta$, *and* $\mathrm{class}(\gamma') = \gamma$.

Proof. Axiom 1 is essentially Propositions 5.3 and 5.5. Axiom 2 is also a consequence of Proposition 5.3. Axiom 3 follows easily from the definitions. Axiom 4 asserts the existence of maps f and g such that $\big(W, V', U', f, g, \theta \circ \Omega^{-1}(\beta)\big)$ is a triangle and such that we have a commutative diagram

$$\begin{array}{ccccccc}
U & \xrightarrow{\alpha} & V & \xrightarrow{\beta} & W & \xrightarrow{\gamma} & \Omega^{-1}(U) \\
\| & & \downarrow{\scriptstyle \mu} & \circlearrowleft & \downarrow{\scriptstyle f} & \circlearrowleft & \| \\
U & \xrightarrow{\mu\alpha} & W' & \xrightarrow{\tau} & V' & \xrightarrow{\rho} & \Omega^{-1}(U) \\
 & & \downarrow{\scriptstyle \nu} & \circlearrowleft & \downarrow{\scriptstyle g} & & \\
V' & \xrightarrow{g} & U' & = & U' & & \\
\downarrow{\scriptstyle \rho} & \circlearrowleft & \downarrow{\scriptstyle \theta} & & \downarrow{\scriptstyle \theta\circ\Omega^{-1}(\beta)} & & \\
\Omega^{-1}(U) & \xrightarrow{\Omega^{-1}(\alpha)} & \Omega^{-1}(V) & \xrightarrow{\Omega^{-1}(\beta)} & \Omega^{-1}(W) & &
\end{array}$$

It is called the Octahedral Axiom because, when we identify U with $\Omega^{-1}(U)$, V with $\Omega^{-1}(V)$, etc., we get a diagram which looks like an octahedron. Four of the faces are the triangles. The axiom says that any two paths with the same beginning and ending points are the same.

A full verification of Axiom 4 is rather complicated, and we will not attempt to include it here. However, we can get the flavor of the axiom by examining the following special case. Namely, imagine that $A \subseteq B \subseteq C$, and $U = A$, $V = B$, $W = B/A$, $W' = C$, $V' = C/A$, $U' = C/B$. We have a commutative diagram with exact rows and columns

$$\begin{array}{ccccccccc}
 & & & & 0 & & 0 & & \\
 & & & & \downarrow & & \downarrow & & \\
0 & \longrightarrow & A & \longrightarrow & B & \longrightarrow & B/A & \longrightarrow & 0 \\
 & & \| & & \downarrow & & \downarrow & & \\
0 & \longrightarrow & A & \longrightarrow & C & \longrightarrow & C/A & \longrightarrow & 0 \\
 & & & & \downarrow & & \downarrow & & \\
 & & & & C/B & = & C/B & & \\
 & & & & \downarrow & & \downarrow & & \\
 & & & & 0 & & 0 & &
\end{array}$$

So the axiom says that $C/B \cong (C/A)/(B/A)$. This is often called the Third Isomorphism Theorem. □

Lemma 5.7 *If $(U, V, W, \alpha, \beta, \gamma)$ is a triangle in ${}_{kG}\mathfrak{stmod}$ and M is in ${}_{kG}\mathfrak{mod}$, then $(U \otimes M, V \otimes M, W \otimes M, \alpha \otimes \mathrm{id}_M, \beta \otimes \mathrm{id}_M, \gamma \otimes \mathrm{id}_M)$ is a triangle in ${}_{kG}\mathfrak{stmod}$.*

Proof. This follows immediately from Theorem 3.3 & Proposition 4.4 (vi) and the definition of a triangle in Theorem 5.6. □

Proposition 5.8 *If $(U, V, W, \alpha, \beta, \gamma)$ is a triangle in ${}_{kG}\mathfrak{stmod}$ and M is in ${}_{kG}\mathfrak{mod}$, then we have long exact sequences*

$$\begin{aligned}
&\underline{\mathrm{Hom}}_{kG}(M, U) \xrightarrow{\alpha_*} \underline{\mathrm{Hom}}_{kG}(M, V) \xrightarrow{\beta_*} \underline{\mathrm{Hom}}_{kG}(M, W) \xrightarrow{\gamma_*} \\
&\xrightarrow{\gamma_*} \mathrm{Ext}^1_{kG}(M, U) \to \mathrm{Ext}^1_{kG}(M, V) \to \mathrm{Ext}^1_{kG}(M, W) \to \\
&\longrightarrow \mathrm{Ext}^2_{kG}(M, U) \to \cdots
\end{aligned} \tag{5.4}$$

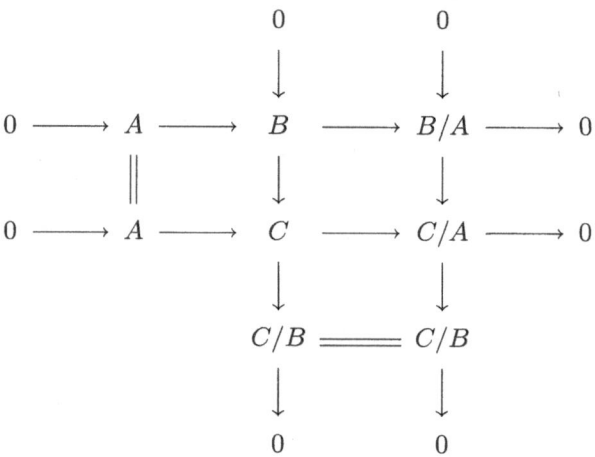

and

$$\underline{\mathrm{Hom}}_{kG}(W,M) \xrightarrow{\beta^*} \underline{\mathrm{Hom}}_{kG}(V,M) \xrightarrow{\alpha^*} \underline{\mathrm{Hom}}_{kG}(U,M) \to$$
$$\longrightarrow \mathrm{Ext}^1_{kG}(W,M) \to \mathrm{Ext}^1_{kG}(V,M) \to \mathrm{Ext}^1_{kG}(U,M) \to$$
$$\longrightarrow \mathrm{Ext}^2_{kG}(W,M) \to \cdots$$

(Note that there are no zeros on the left ends.)

Proof. Recall that for $n > 0$ (see Theorem 5.1 and Proposition 5.2) $\mathrm{Ext}^n_{kG}(M,U) \cong \underline{\mathrm{Hom}}_{kG}(M,\Omega^{-n}(U))$ and $\mathrm{Ext}^n_{kG}(W,M) \cong \underline{\mathrm{Hom}}_{kG}(\Omega^n(W),M)$.

We show that there is a long exact sequence

$$\cdots \xrightarrow{\Omega^2(\gamma)_*} \underline{\mathrm{Hom}}_{kG}(M,\Omega(U)) \xrightarrow{\Omega(\alpha)_*} \underline{\mathrm{Hom}}_{kG}(M,\Omega(V)) \xrightarrow{\Omega(\beta)_*} \underline{\mathrm{Hom}}_{kG}(M,\Omega(W))$$
$$\xrightarrow{\Omega(\gamma)_*} \underline{\mathrm{Hom}}_{kG}(M,U) \xrightarrow{\alpha_*} \underline{\mathrm{Hom}}_{kG}(M,V) \xrightarrow{\beta_*} \underline{\mathrm{Hom}}_{kG}(M,W) \xrightarrow{\gamma_*}$$
$$\xrightarrow{\gamma_*} \underline{\mathrm{Hom}}_{kG}(M,\Omega^{-1}(U)) \xrightarrow{\Omega^{-1}(\alpha)_*} \underline{\mathrm{Hom}}_{kG}(M,\Omega^{-1}(V)) \xrightarrow{\Omega^{-1}(\beta)_*} \cdots$$

which extends (5.4) to the left.

By Axiom 2 it is enough to show that

$$\underline{\mathrm{Hom}}_{kG}(M,\Omega^{-n}(U)) \xrightarrow{\Omega^{-n}(\alpha)_*} \underline{\mathrm{Hom}}_{kG}(M,\Omega^{-n}(V)) \xrightarrow{\Omega^{-n}(\beta)_*} \underline{\mathrm{Hom}}_{kG}(M,\Omega^{-n}(W))$$

is exact. Surely $\mathrm{im}\,\Omega^{-n}(\alpha)_* \subseteq \ker \Omega^{-n}(\beta)_*$. Let $\zeta \in \ker \Omega^{-n}(\beta)_*$. The rows of the following diagram define triangles

$$\begin{array}{ccccccc}
\Omega^n(M) & \longrightarrow & 0 & \longrightarrow & \Omega^{n-1}(M) & \xrightarrow{-\mathrm{id}_{\Omega^{n-1}(M)}} & \Omega^{n-1}(M) \\
\downarrow{\scriptstyle \Omega^n(\zeta)} & & \downarrow & & & & \downarrow{\scriptstyle \Omega^{n-1}(\zeta)} \\
V & \xrightarrow{\beta} & W & \xrightarrow{\gamma} & \Omega^{-1}(U) & \xrightarrow{-\Omega^{-1}(\alpha)} & \Omega^{-1}(V)
\end{array}$$

By Axiom 3 there exists $\eta : \Omega^{n-1}(M) \to \Omega^{-1}(U)$ with $\Omega^{-1}(\alpha) \circ \eta = \Omega^{n-1}(\zeta)$. Hence $\zeta = \Omega^{-n}(\alpha) \circ \Omega^{-n+1}(\eta) \in \mathrm{im}\,\Omega^{-n}(\alpha)_*$.

The exactness for the other sequence follows dually. □

Corollary 5.9 *If* $(U,V,W,\alpha,\beta,0)$ *is a triangle in* ${}_{kG}\mathfrak{stmod}$, *then* $U \oplus W \cong V \oplus (\mathrm{proj})$ *in* ${}_{kG}\mathfrak{mod}$.

Proof. We must show that the short exact sequence from Theorem 5.6

$$0 \longrightarrow U \xrightarrow{\alpha'} V \oplus (\text{proj}) \xrightarrow{\beta'} W \longrightarrow 0 \tag{5.5}$$

splits. The usual long exact cohomology sequence for (5.5) reads

$$\cdots \longrightarrow \operatorname{Hom}_{kG}\big(W, V \oplus (\text{proj})\big) \xrightarrow{\beta'_*} \operatorname{Hom}_{kG}(W, W) \xrightarrow{\delta} \operatorname{Ext}^1_{kG}(W, U) \xrightarrow{\alpha'_*}$$
$$\xrightarrow{\alpha'_*} \operatorname{Ext}^1_{kG}\big(W, V \oplus (\text{proj})\big) \to \cdots$$

The composition of α'_* with the canonical isomorphism $\operatorname{Ext}^1_{kG}\big(W, V \oplus (\text{proj})\big) \xrightarrow{\cong} \operatorname{Ext}^1_{kG}(W, V)$ is the morphism which occurs in the long exact sequence (5.4) for $M = W$. Proposition 5.8 with $\gamma = 0$ and $M = W$ shows that α'_* is injective. Hence $\delta = 0$ and β'_* is surjective. Now $\theta \in \operatorname{Hom}_{kG}\big(W, V \oplus (\text{proj})\big)$ with $\operatorname{id}_W = \beta'_* \circ \theta = \beta\theta$ is the required splitting map. □

6 Products in cohomology

In this section we define the cup product in group cohomology and show some of its properties. The existence of the tensor product operation permits a host of variations on the classical definition. The remarkable thing is that they are all the same. We begin with a basic result on the tensor product of complexes.

Suppose that the complexes C_* and D_* are both either bounded above or bounded below. That is, either $C_n = 0$ and $D_n = 0$ for all n sufficiently large or both $C_{-n} = 0$ and $D_{-n} = 0$ for all n sufficiently large. The tensor product $C_* \otimes D_*$ is defined to be the complex with

$$(C_* \otimes D_*)_n := \bigoplus_{i+j=n} C_i \otimes D_j$$

and with boundary homomorphisms given by

$$\partial(x \otimes y) := \partial x \otimes y + (-1)^{\deg x} x \otimes \partial y$$

if x is a homogeneous element of degree $\deg x$, that is, if $x \in C_{\deg x}$.

Exercise 6.1 Show that $\partial\partial = 0$.

Exercise 6.2 Let

$$0 \to A \xrightarrow{\alpha} B \xrightarrow{\beta} C \to 0 \quad \text{and} \quad 0 \to D \xrightarrow{\gamma} E \xrightarrow{\delta} F \to 0$$

be two short exact sequences, and define the two complexes $\mathcal{C}_*$ and $\mathcal{D}_*$ by

$$\begin{aligned} \mathcal{C}_* &: \quad 0 \to B \xrightarrow{\beta} C \to 0 \quad \text{with } C \text{ in degree } 0, \\ \mathcal{D}_* &: \quad 0 \to E \xrightarrow{\delta} F \to 0 \quad \text{with } F \text{ in degree } 0. \end{aligned}$$

Then we have

$$\underbrace{0 \to A \otimes D \xrightarrow{\alpha\otimes\gamma} B \otimes E \xrightarrow{\binom{-1\otimes\delta}{\beta\otimes 1}} \begin{matrix} B \otimes F \\ \oplus \\ C \otimes E \end{matrix} \xrightarrow{(\beta\otimes 1 \;\; 1\otimes\delta)} C \otimes F \to 0}_{\mathcal{C}_* \otimes \mathcal{D}_*} \tag{6.1}$$

Show that the sequence (6.1) is exact. Notice that $H_*(\mathcal{C}_*) = H_1(\mathcal{C}_*) \cong A$, $H_*(\mathcal{D}_*) = H_1(\mathcal{D}_*) \cong D$, $H_*(\mathcal{C}_* \otimes \mathcal{D}_*) = H_2(\mathcal{C}_* \otimes \mathcal{D}_*) \cong A \otimes D$.

The Künneth Theorem which follows is stated very differently from what would be found in most texts. Our statement and proof rely heavily on the fact that we are taking our tensor products over a field. Otherwise a Tor term would be involved in the homology. (See [HS] for a more thorough treatment.) A key point is that the isomorphism in Theorem 6.1 is an isomorphism of kG-modules.

Theorem 6.1 *Let C_* and D_* be complexes in ${}_{kG}\mathfrak{mod}$ both of which are either bounded above or bounded below. Then for any integer n we have that*

$$H_n(C_* \otimes D_*) \cong \bigoplus_{i+j=n} H_i(C_*) \otimes H_j(D_*)$$

as kG-modules.

Proof. The complex C_* is $\cdots \to C_n \xrightarrow{\partial_n} C_{n-1} \to \cdots$. Define $Z_n := \ker \partial_n \subseteq C_n$ and $B_{n-1} := \operatorname{im} \partial_n \subseteq Z_{n-1} \subseteq C_{n-1}$, so that $B_{n-1} \cong C_n/Z_n$. We consider Z_* and B_* as complexes with zero differentials. So we have an exact sequence of complexes

$$0 \to Z_* \to C_* \to B_{*-1} \to 0 \tag{6.2}$$

where the chain map $Z_* \to C_*$ is the inclusion and $C_* \to B_{*-1}$ is the map of complexes induced by ∂_*. By tensoring the sequence (6.2) with D_*, we get the exact sequence

$$0 \to Z_* \otimes D_* \to C_* \otimes D_* \to B_{*-1} \otimes D_* \to 0$$

whose long exact homology sequence reads

$$\cdots \to H_n(Z_* \otimes D_*) \to H_n(C_* \otimes D_*) \to H_n(B_{*-1} \otimes D_*) \xrightarrow{\delta} H_{n-1}(Z_* \otimes D_*) \to \cdots \tag{6.3}$$

Since the differential in Z_* is zero, we have that $Z_* \otimes D_* = \bigoplus_i Z_i \otimes D_*$ as a complex and $H_n(Z_* \otimes D_*) \cong \bigoplus_{i+j=n} Z_i \otimes H_j(D_*)$. By the same argument we have isomorphisms $H_n(B_{*-1} \otimes D_*) \cong \bigoplus_{i+j=n-1} B_i \otimes H_j(D_*)$.

Our claim is that the connecting homomorphism δ in the long exact homology sequence (6.3) is injective. To prove this consider the diagram

$$\begin{array}{ccccccccc}
0 & \longrightarrow & \bigoplus_{i+j=n} Z_i \otimes D_j & \longrightarrow & \bigoplus_{i+j=n} C_i \otimes D_j & \xrightarrow{\partial\otimes 1} & \bigoplus_{i+j=n-1} B_i \otimes D_j & \longrightarrow & 0 \\
 & & \downarrow{\scriptstyle \bigoplus_{i+j=n} (-1)^i 1\otimes\partial} & & \downarrow{\scriptstyle \beta} & & \downarrow{\scriptstyle \bigoplus_{i+j=n-1} (-1)^i 1\otimes\partial} & & \\
0 & \longrightarrow & \bigoplus_{i+j=n-1} Z_i \otimes D_j & \longrightarrow & \bigoplus_{i+j=n-1} C_i \otimes D_j & \longrightarrow & \bigoplus_{i+j=n-2} B_i \otimes D_j & \longrightarrow & 0
\end{array}$$

where $\beta = \bigoplus_{i+j=n} (\partial \otimes 1 + (-1)^i 1 \otimes \partial)$. Let $x = b \otimes d \in B_i \otimes H_j(D_*) \hookrightarrow H_n(B_{*-1} \otimes D_*)$ and let $d' \in \ker \partial \subseteq D_j$ be a representative for d. To construct δx one can first take a preimage $b' \otimes d' \in C_i \otimes D_j$ of $b \otimes d'$. Then $\beta(b' \otimes d') = b \otimes d' \in C_i \otimes D_j$ lies actually in $Z_i \otimes D_j$. Its homology class $b \otimes d \in Z_i \otimes H_j(D_*)$ is δx. In other words,

$$\bigoplus_{i+j=n-1} B_i \otimes H_j(D_*) \cong H_n(B_{*-1} \otimes D_*) \xrightarrow{\delta} H_{n-1}(Z_* \otimes D_*) \cong \bigoplus_{i+j=n-1} Z_i \otimes H_j(D_*)$$

is just the map $\bigoplus_{i+j=n-1} (\text{inclusion } B_i \hookrightarrow Z_i) \otimes \mathrm{id}_{H_j(D_*)}$, which is evidently injective.

From (6.3) together with the fact that δ is injective, we get short exact sequences as in following commutative diagram:

$$\begin{array}{ccccccc}
0 \longrightarrow & H_n(B_* \otimes D_*) & \xrightarrow{\delta} & H_n(Z_* \otimes D_*) & \longrightarrow & H_n(C_* \otimes D_*) & \longrightarrow 0 \\
 & \wr\Vert & & \wr\Vert & & & \\
0 \longrightarrow & \bigoplus_{i+j=n} B_i \otimes H_j(D_*) & \hookrightarrow & \bigoplus_{i+j=n} Z_i \otimes H_j(D_*) & \longrightarrow & \bigoplus_{i+j=n} (Z_i/B_i) \otimes H_j(D_*) & \longrightarrow 0 \\
 & & & & & \Vert & \\
 & & & & & \bigoplus_{i+j=n} H_i(C_*) \otimes H_j(D_*) &
\end{array}$$

Hence $H_n(C_* \otimes D_*) \cong \bigoplus_{i+j=n} H_i(C_*) \otimes H_j(D_*)$. □

Corollary 6.2 *If* $P_* \xrightarrow{\varepsilon} M$ *and* $Q_* \xrightarrow{\theta} N$ *are projective resolutions, then* $P_* \otimes Q_* \xrightarrow{\varepsilon\otimes\theta} M \otimes N$ *is a projective resolution.*

Next we review the representation of $\operatorname{Ext}^n_{kG}(M,N)$ as extension classes. Here M,N are in ${}_{kG}\mathfrak{mod}$, and n is a nonnegative integer. Let $\mathcal{U}^n(M,N)$ be the class of all exact sequences in ${}_{kG}\mathfrak{mod}$ of the form

$$E: \quad 0 \longrightarrow N \longrightarrow B_{n-1} \longrightarrow \cdots \longrightarrow B_0 \longrightarrow M \longrightarrow 0.$$

Define a relation $\xrightarrow{\simeq}$ on $\mathcal{U}^n(M,N)$ by $E_1 \xrightarrow{\simeq} E_2$ if there is a chain map θ_*

$$\begin{array}{cccccccccccc}
E_1: & 0 & \longrightarrow & N & \longrightarrow & B_{n-1} & \longrightarrow & \cdots & \longrightarrow & B_0 & \longrightarrow & M \longrightarrow 0 \\
 & & & \| & & \downarrow{\scriptstyle\theta_{n-1}} & & & & \downarrow{\scriptstyle\theta_0} & & \| \\
E_2: & 0 & \longrightarrow & N & \longrightarrow & C_{n-1} & \longrightarrow & \cdots & \longrightarrow & C_0 & \longrightarrow & M \longrightarrow 0
\end{array}$$

That is, $\theta_n = \mathrm{id}_N$ and $\theta_{-1} = \mathrm{id}_M$. The relation $\xrightarrow{\simeq}$ is not an equivalence relation because it's not symmetric. Let $\sim$ be the minimal equivalence relation on $\mathcal{U}^n(M,N)$ containing $\xrightarrow{\simeq}$. In other words $E_1 \sim E_2$ provided there exists a chain $F_0,\dots,F_f$ in $\mathcal{U}^n(M,N)$ with $E_1 = F_0$, $E_2 = F_f$, and for each $i = 1,\dots,f$ either $F_{i-1} \xrightarrow{\simeq} F_i$ or $F_i \xrightarrow{\simeq} F_{i-1}$.

Theorem 6.3 *Let M,N be in ${}_{kG}\mathfrak{mod}$ and n a nonnegative integer. Then there is a bijection*

$$\operatorname{Ext}^n_{kG}(M,N) \underset{\theta}{\overset{\psi}{\rightleftarrows}} \mathcal{U}^n(M,N)/\!\sim .$$

Proof. (Sketch) Let $P_* \overset{\varepsilon}{\twoheadrightarrow} M$ be a projective resolution.

Given $E \in \mathcal{U}^n(M,N)$ we get a chain map μ_*

$$\begin{array}{ccccccccc}
 & \cdots \to P_{n+1} & \xrightarrow{\partial} & P_n & \longrightarrow & P_{n-1} \to \cdots \to P_0 & \longrightarrow & M \longrightarrow 0 \\
 & \downarrow{\scriptstyle\mu_{n+1}} & & \downarrow{\scriptstyle\mu_n} & & \downarrow{\scriptstyle\mu_{n-1}} \qquad \downarrow{\scriptstyle\mu_0} & & \| \\
E: & 0 & \longrightarrow & N & \longrightarrow & B_{n-1} \to \cdots \to B_0 & \longrightarrow & M \longrightarrow 0
\end{array}$$

Since $\mu_n : P_n \to N$ is a cocycle ($\mu_n \circ \partial = 0$), we can look at the class $\mathrm{class}(\mu_n) \in \operatorname{Ext}^n_{kG}(M,N)$. One has now to verify that $\theta : \mathrm{class}(E) \mapsto \mathrm{class}(\mu_n)$ is well-defined.

Suppose we are given $\zeta \in \operatorname{Ext}^n_{kG}(M,N)$. Then ζ is represented by a cocycle $\hat{\zeta} : P_n \to N$. We have a commutative diagram

$$\begin{array}{cccccccccc}
 & \cdots \to P_{n+1} & \longrightarrow & P_n & \xrightarrow{\partial} & P_{n-1} & \longrightarrow & P_{n-2} \to \cdots \to P_0 & \longrightarrow & M \longrightarrow 0 \\
 & \downarrow & & {\scriptstyle\hat{\zeta}}\downarrow & \text{pushout} & \downarrow & & \| \qquad\qquad \| & & \| \\
E: & 0 & \longrightarrow & N & \longrightarrow & B & \longrightarrow & P_{n-2} \to \cdots \to P_0 & \longrightarrow & M \longrightarrow 0
\end{array}$$

where $B := N \oplus P_{n-1}/\{(\hat{\zeta}(x), -\partial x) \mid x \in P_n\}$ is the pushout of the diagram defined by $\hat{\zeta}$ and ∂. The lower row E is then exact. So it is an element in $\mathcal{U}^n(M,N)$. One has now to show that $\psi : \zeta \mapsto \text{class}(E)$ is well-defined and that θ and ψ are inverses of each other. □

The products in cohomology come in two sorts. There are *outer products*, which make no assumption on the factors. For the *inner products* some sort of composition must be guaranteed. We begin with a list of possible outer products.

Suppose that $\zeta \in \text{Ext}^m_{kG}(M,N)$ and $\gamma \in \text{Ext}^n_{kG}(M',N')$.

(1) *Products by projective resolutions.* Suppose that $P_* \overset{\varepsilon}{\twoheadrightarrow} M$ and $P'_* \overset{\varepsilon'}{\twoheadrightarrow} M'$ are projective resolutions. Then ζ and γ are represented by $\tilde{\zeta} : P_m \to N$ and $\tilde{\gamma} : P'_n \to N'$, respectively. Define $\zeta\gamma$ to be the class of

$$\tilde{\zeta} \otimes \tilde{\gamma} : (P_* \otimes P'_*)_{m+n} \twoheadrightarrow P_m \otimes P'_n \to N \otimes N'$$

for the projective resolution $P_* \otimes P'_* \xrightarrow{\varepsilon\otimes\varepsilon'} M \otimes M'$.

(1′) Similarly if $N \overset{\theta}{\hookrightarrow} I_*$ and $N' \overset{\theta'}{\hookrightarrow} I'_*$ are injective resolutions, and ζ, γ are represented by $\hat{\zeta} : M \to I_{-m}$, $\hat{\gamma} : M' \to I'_{-n}$, then let $\zeta\gamma$ be the class of

$$\hat{\zeta} \otimes \hat{\gamma} : M \otimes M' \to I_{-m} \otimes I'_{-n} \hookrightarrow (I_* \otimes I'_*)_{-m-n}.$$

(2) *Products of maps in the stable category.* Here we assume that $m > 0$, $n > 0$. Let ζ and γ be represented by $\hat{\zeta} \in \underline{\text{Hom}}_{kG}(\Omega^m(M), N)$ and $\hat{\gamma} \in \underline{\text{Hom}}_{kG}(\Omega^n(M'), N')$. Then let $\zeta\gamma$ be the class of $\hat{\zeta} \otimes \hat{\gamma} \in \underline{\text{Hom}}_{kG}(\Omega^m(M) \otimes \Omega^n(M'), N \otimes N') \cong \underline{\text{Hom}}_{kG}(\Omega^{m+n}(M \otimes M'), N \otimes N') \cong \text{Ext}^{m+n}_{kG}(M \otimes M', N \otimes N')$.

(3) *Products by the tensor product of complexes.* Suppose that ζ and γ are represented by

$$E_\zeta : \; 0 \longrightarrow N \longrightarrow A_{m-1} \longrightarrow \cdots \longrightarrow A_0 \longrightarrow M \longrightarrow 0$$

and

$$E_\gamma : \; 0 \longrightarrow N' \longrightarrow B_{n-1} \longrightarrow \cdots \longrightarrow B_0 \longrightarrow M' \longrightarrow 0.$$

Let C_* be

$$\begin{array}{ccccccccccc} 0 & \longrightarrow & A_{m-1} & \longrightarrow & \cdots & \longrightarrow & A_0 & \longrightarrow & M & \longrightarrow & 0 \\ & & \| & & & & \| & & \| & & \\ 0 & \longrightarrow & C_m & \longrightarrow & \cdots & \longrightarrow & C_1 & \longrightarrow & C_0 & \longrightarrow & 0 \end{array}$$

and D_* be

$$\begin{array}{ccccccccccc} 0 & \longrightarrow & B_{n-1} & \longrightarrow & \cdots & \longrightarrow & B_0 & \longrightarrow & M' & \longrightarrow & 0 \\ & & \| & & & & \| & & \| & & \\ 0 & \longrightarrow & D_n & \longrightarrow & \cdots & \longrightarrow & D_1 & \longrightarrow & D_0 & \longrightarrow & 0. \end{array}$$

Then $H_*(C_* \otimes D_*) = H_{m+n}(C_* \otimes D_*) \cong N \otimes N'$. So we have an exact sequence

$$\begin{array}{l} 0 \longrightarrow N \otimes N' \longrightarrow (C_* \otimes D_*)_{m+n} \longrightarrow \cdots \longrightarrow (C_* \otimes D_*)_0 \longrightarrow 0. \\ \hphantom{0 \longrightarrow N \otimes N' \longrightarrow (C_* \otimes D_*)_{m+n} \longrightarrow \cdots \longrightarrow (C_*} \cong \\ \hphantom{0 \longrightarrow N \otimes N' \longrightarrow (C_* \otimes D_*)_{m+n} \longrightarrow \cdots \longrightarrow} M \otimes M' \end{array}$$

Define $\zeta\gamma$ to be its class in $\mathrm{Ext}_{kG}^{m+n}(M \otimes M', N \otimes N')$.

(3′) Let C_* be defined by

$$\begin{array}{ccccccccccc} 0 & \longrightarrow & N & \longrightarrow & A_{m-1} & \longrightarrow & \cdots & \longrightarrow & A_0 & \longrightarrow & 0 \\ & & \| & & \| & & & & \| & & \\ 0 & \longrightarrow & C_0 & \longrightarrow & C_{-1} & \longrightarrow & \cdots & \longrightarrow & C_{-m} & \longrightarrow & 0 \end{array}$$

and let D_* be given by

$$\begin{array}{ccccccccccc} 0 & \longrightarrow & N' & \longrightarrow & B_{n-1} & \longrightarrow & \cdots & \longrightarrow & B_0 & \longrightarrow & 0 \\ & & \| & & \| & & & & \| & & \\ 0 & \longrightarrow & D_0 & \longrightarrow & D_{-1} & \longrightarrow & \cdots & \longrightarrow & D_{-n} & \longrightarrow & 0. \end{array}$$

Then $H_*(C_* \otimes D_*) = H_{-m-n}(C_* \otimes D_*) \cong M \otimes M'$. So we have an exact sequence

$$\begin{array}{l} 0 \longrightarrow (C_* \otimes D_*)_0 \longrightarrow \cdots \longrightarrow (C_* \otimes D_*)_{-m-n} \longrightarrow M \otimes M' \longrightarrow 0 \\ \hphantom{0 \longrightarrow (C_*} \cong \\ \hphantom{0 \longrightarrow (C} N \otimes N' \end{array}$$

whose class is defined to be $\zeta\gamma$.

(4) *Yoneda splice products.* For ζ and γ represented as in (3), let $\zeta\gamma$ be the class of the sequence $(E_\zeta \otimes N') \circ (M \otimes E_\gamma)$:

$$\begin{array}{l} 0 \to N \otimes N' \to A_{m-1} \otimes N' \to \cdots \to A_0 \otimes N' \longrightarrow M \otimes B_{n-1} \to \cdots \\ \hphantom{0 \to N \otimes N' \to A_{m-1} \otimes N' \to \cdots \to A_0} \searrow \quad \nearrow \\ \hphantom{0 \to N \otimes N' \to A_{m-1} \otimes N' \to \cdots \to A_0 \otimes} M \otimes N' \\ \hphantom{0 \to N \otimes N' \to A_{m-1} \otimes N' \to} \cdots \to M \otimes B_0 \to M \otimes M' \to 0. \end{array}$$

The general principle here is that almost any reasonable product in cohomology is the right one. For outer products the principle is expressed in the following theorem.

Theorem 6.4 *All of the outer products* (1), (1′), (2), (3), (3′), *and* (4) *coincide.*

Proof. We prove the theorem by considering a single (though large) commutative diagram. Most of the commutativity in the diagram follows from the fact that the product of two chain maps is again a chain map. As input for our main diagram we have the following commutative diagram. The notation is taken from the definition of the products.

We may suppose that the chosen projective and injective resolutions are minimal.

$$
\begin{array}{ccccccccccc}
\to P_{m+1} & \to & P_m & \longrightarrow & P_{m-1} & \to \cdots \to & P_0 & \xrightarrow{\varepsilon} & M & \longrightarrow & 0 \\
\downarrow & & \downarrow{\scriptstyle \nu_m} & & \| & & \| & & \| & & \\
0 & \to & \Omega^m(M) & \to & P_{m-1} & \to \cdots \to & P_0 & \longrightarrow & M & \longrightarrow & 0 \\
 & & \downarrow{\scriptstyle \mu_m} & & \downarrow{\scriptstyle \mu_{m-1}} & & \downarrow{\scriptstyle \mu_0} & & \| & & \\
E_\zeta : 0 & \longrightarrow & N & \xrightarrow{\iota} & A_{m-1} & \to \cdots \to & A_0 & \xrightarrow{\pi} & M & \longrightarrow & 0 \\
 & & \| & & \downarrow{\scriptstyle \sigma_0} & & \downarrow{\scriptstyle \sigma_{-m+1}} & & \downarrow{\scriptstyle \sigma_{-m}} & & \\
0 & \longrightarrow & N & \longrightarrow & I_0 & \to \cdots \to & I_{-m+1} & \to & \Omega^{-m}(M) & \to & 0 \\
 & & \| & & \| & & \| & & \downarrow{\scriptstyle \tau_{-m}} & & \downarrow \\
0 & \longrightarrow & N & \xrightarrow{\theta} & I_0 & \to \cdots \to & I_{-m+1} & \longrightarrow & I_{-m} & \to & I_{-m-1} \to
\end{array}
$$

The chain maps μ_*, ν_*, σ_*, and τ_* are constructed in the obvious way. The maps $\mu_m \circ \nu_m : P_m \to N$ and $\tau_{-m} \circ \sigma_{-m} : M \to I_{-m}$ are cocycles representing $\zeta = \text{class}(E_\zeta)$.

There is a similar diagram for the cohomology element $\gamma \in \text{Ext}^n_{kG}(M', N')$.

It is a substantial check to see that the diagram on the next page commutes. Start with (4). For (3) we take E_ζ and cut off M and take E_γ and cut off M'. Their tensor product is then a resolution of $M \otimes M'$. For (2) we take the tensor product of the resolutions in the second rows of the preparatory diagrams. Finally (1) is just $P_* \otimes P'_*$.

Actually the commutativity in rows (1), (2), and (3) and also in rows (1′), (2′), and (3′) follows from the fact that each chain map connecting these rows is a tensor product of two other chain maps. The commutativity of (3), (4), and (3′) must be checked more carefully.

$$
\begin{array}{lccccccccccc}
(1) & \cdots\to\bigoplus\limits_{i+j=m+n} P_i\otimes P'_j & \longrightarrow & \bigoplus\limits_{i+j=m+n-1} P_i\otimes P'_j & \to\cdots\to & \bigoplus\limits_{i+j=n} P_i\otimes P'_j & \longrightarrow & \bigoplus\limits_{i+j=n-1} P_i\otimes P'_j & \to\cdots\to & P_0\otimes P'_0 & \longrightarrow & M\otimes M'\to 0 \\
 & \Big\downarrow{\scriptstyle (\nu_*\otimes\nu'_*)_{m+n}} & & \Big\downarrow{\scriptstyle (\nu_*\otimes\nu'_*)_{m+n-1}} & & \Big\downarrow{\scriptstyle (\nu_*\otimes\nu'_*)_{n}} & & \Big\downarrow{\scriptstyle (\nu_*\otimes\nu'_*)_{n-1}} & & \Vert & & \Vert \\
(2) & 0\to\begin{matrix}\Omega^m(M)\\ \otimes\,\Omega^n(M')\end{matrix} & \to & \begin{matrix}P_{m-1}\otimes\Omega^n(M')\\ \oplus\\ \Omega^m(M)\otimes P'_{n-1}\end{matrix} & \to\cdots\to & \begin{matrix}P_0\otimes\Omega^n(M')\\ \oplus\\ \text{(other)}\end{matrix} & \longrightarrow & \begin{matrix}P_0\otimes P'_{n-1}\\ \oplus\\ \text{(other)}\end{matrix} & \to\cdots\to & P_0\otimes P'_0 & \longrightarrow & M\otimes M'\to 0 \\
 & \Big\downarrow{\scriptstyle \mu_m\otimes\mu'_n} & & \Big\downarrow & & \Big\downarrow{\scriptstyle \mu_0\otimes\mu'_n+\cdots} & & \Big\downarrow & & \Big\downarrow{\scriptstyle \mu_0\otimes\mu'_0} & & \Vert \\
(3) & 0\to N\otimes N' & \xrightarrow{\binom{\iota\otimes 1}{1\otimes\iota'}} & \begin{matrix}A_{m-1}\otimes N'\\ \oplus\\ N\otimes B_{n-1}\end{matrix} & \to\cdots\to & \begin{matrix}A_0\otimes N'\\ \oplus\\ \text{(other)}\end{matrix} & \xrightarrow{\substack{1\otimes\iota'\\ +\cdots}} & \begin{matrix}A_0\otimes B_{n-1}\\ \oplus\\ \text{(other)}\end{matrix} & \to\cdots\to & A_0\otimes B_0 & \xrightarrow{\pi\otimes\pi'} & M\otimes M'\to 0 \\
 & \Vert & & \Big\downarrow{\scriptstyle \text{projection}} & & \Big\downarrow{\scriptstyle \text{projection}} & & \Big\downarrow{\scriptstyle \pi\otimes 1} & & \Big\downarrow{\scriptstyle \pi\otimes 1} & & \Vert \\
(4) & 0\to N\otimes N' & \xrightarrow{\iota\otimes 1} & A_{m-1}\otimes N' & \to\cdots\to & A_0\otimes N' & \xrightarrow{\pi\otimes\iota'} & M\otimes B_{n-1} & \to\cdots\to & M\otimes B_0 & \xrightarrow{1\otimes\pi'} & M\otimes M'\to 0 \\
 & \Vert & & \Big\downarrow{\scriptstyle 1\otimes\iota'} & & \Big\downarrow{\scriptstyle 1\otimes\iota'} & & \Big\downarrow{\scriptstyle \text{inclusion}} & & \Big\downarrow{\scriptstyle \text{inclusion}} & & \Vert \\
(3') & 0\to N\otimes N' & \xrightarrow{\iota\otimes\iota'} & A_{m-1}\otimes B_{n-1} & \to\cdots\to & \begin{matrix}A_0\otimes B_{n-1}\\ \oplus\\ \text{(other)}\end{matrix} & \xrightarrow{\substack{\pi\otimes 1\\ +\cdots}} & \begin{matrix}M\otimes B_{n-1}\\ \oplus\\ \text{(other)}\end{matrix} & \to\cdots\to & \begin{matrix}M\otimes B_0\\ \oplus\\ A_0\otimes M'\end{matrix} & \xrightarrow{\substack{1\otimes\pi'\\ +\pi\otimes 1}} & M\otimes M'\to 0 \\
 & \Vert & & \Big\downarrow{\scriptstyle \sigma_0\otimes\sigma'_0} & & \Big\downarrow & & \Big\downarrow{\scriptstyle \sigma_{-m}\otimes\sigma'_0+\cdots} & & \Big\downarrow & & \Big\downarrow{\scriptstyle \sigma_{-m}\otimes\sigma'_{-n}} \\
(2') & 0\to N\otimes N' & \longrightarrow & I_0\otimes I'_0 & \to\cdots\to & \begin{matrix}I_{-m+1}\otimes I'_0\\ \oplus\\ \text{(other)}\end{matrix} & \longrightarrow & \begin{matrix}\Omega^{-m}(N)\otimes I'_0\\ \oplus\\ \text{(other)}\end{matrix} & \to\cdots\to & \begin{matrix}\Omega^{-m}(N)\otimes I'_{-n+1}\\ \oplus\\ I_{-m+1}\otimes\Omega^{-n}(N')\end{matrix} & \to & \begin{matrix}\Omega^{-m}(N)\\ \otimes\,\Omega^{-n}(N')\end{matrix}\to 0 \\
 & \Vert & & \Vert & & \Big\downarrow{\scriptstyle (\tau_*\otimes\tau'_*)_{-m+1}} & & \Big\downarrow{\scriptstyle (\tau_*\otimes\tau'_*)_{-m}} & & \Big\downarrow{\scriptstyle (\tau_*\otimes\tau'_*)_{-m-n+1}} & & \Big\downarrow{\scriptstyle (\tau_*\otimes\tau'_*)_{-m-n}} \\
(1') & 0\to N\otimes N' & \longrightarrow & I_0\otimes I'_0 & \to\cdots\to & \bigoplus\limits_{i+j=-m+1} I_i\otimes I'_j & \longrightarrow & \bigoplus\limits_{i+j=-m} I_i\otimes I'_j & \to\cdots\to & \bigoplus\limits_{i+j=-m-n+1} I_i\otimes I'_j & \longrightarrow & \bigoplus\limits_{i+j=-m-n} I_i\otimes I'_j\to\cdots
\end{array}
$$

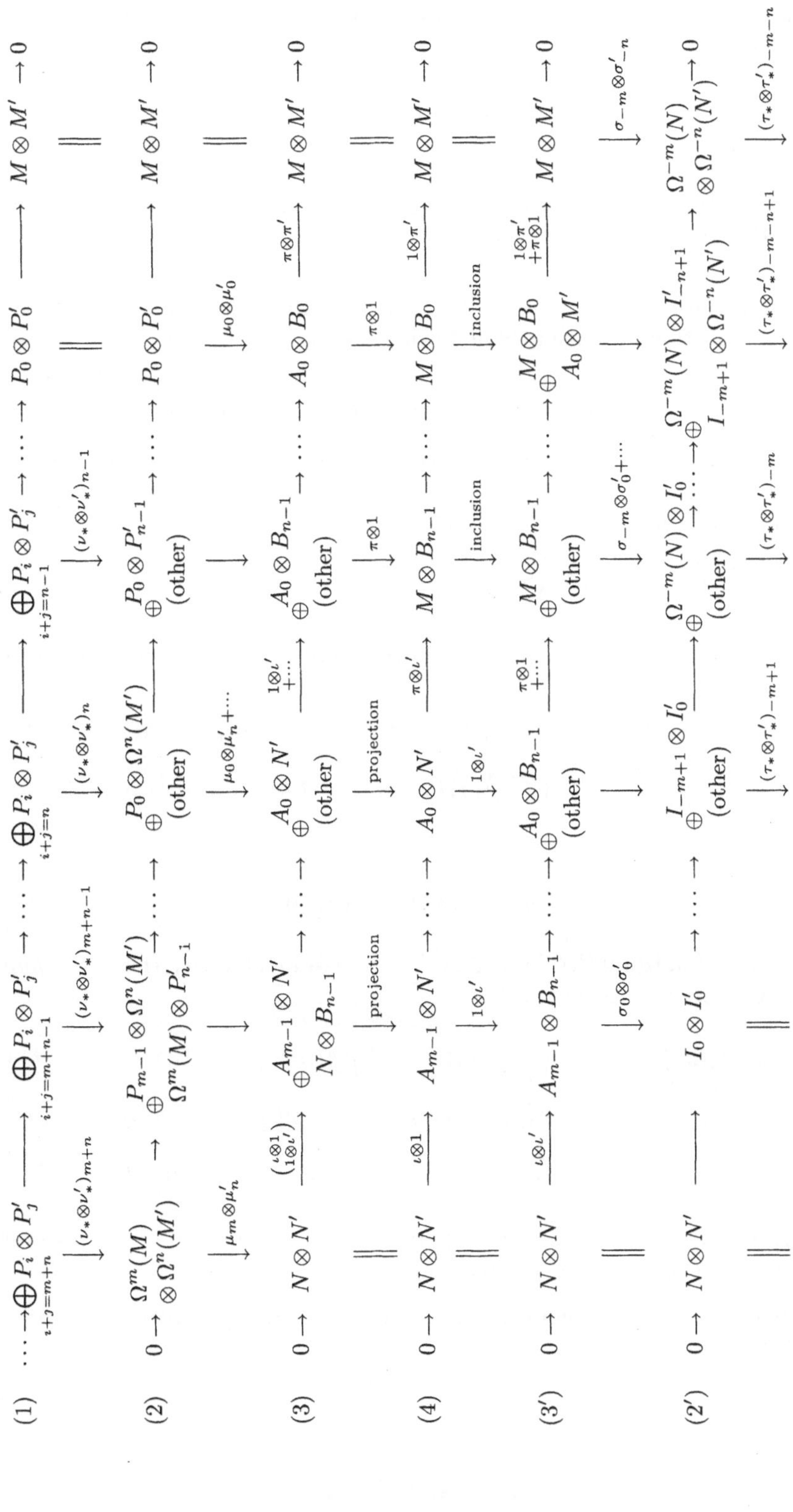

The vertical map $(\nu_* \otimes \nu'_*)_{m+n}$ in the upper left corner in the diagram is the composition

$$\bigoplus_{i+j=m+n} P_i \otimes P'_j \xrightarrow{\text{projection}} P_m \otimes P'_n \xrightarrow{\nu_m \otimes \nu'_n} \Omega^m(M) \otimes \Omega^n(M'),$$

which factors through $\Omega^{m+n}(M \otimes M') \oplus (\text{proj})$. An analogous statement can be made for the lower left corner of the diagram. □

Now we turn our attention to the inner products. In order to make sense of one of the definitions we need the following lemma.

Lemma 6.5 *Let A, B, and C be in ${}_{kG}\mathfrak{mod}$. Then there are natural isomorphisms*

$$\mathrm{Hom}_{kG}(A \otimes B, C) \cong \mathrm{Hom}_{kG}\big(A, \mathrm{Hom}(B, C)\big) \cong \mathrm{Hom}_{kG}(A, B^* \otimes C), \qquad (6.4)$$

and, more generally, for any nonnegative integer n

$$\mathrm{Ext}^n_{kG}(A \otimes B, C) \cong \mathrm{Ext}^n_{kG}\big(A, \mathrm{Hom}(B, C)\big) \cong \mathrm{Ext}^n_{kG}(A, B^* \otimes C).$$

Proof. In fact, all three vector spaces in (6.4) are isomorphic to the G-invariants (i. e. G-fixed points) of $A^* \otimes B^* \otimes C$.

More explicitly, define $\mathrm{Hom}_{kG}(A \otimes B, C) \underset{\theta}{\overset{\psi}{\rightleftarrows}} \mathrm{Hom}_{kG}\big(A, \mathrm{Hom}(B, C)\big)$ by

$$\big(\psi(f)(a)\big)(b) := f(a \otimes b)$$

for $f \in \mathrm{Hom}_{kG}(A \otimes B, C)$, $a \in A$, $b \in B$ and

$$\theta(g)(a \otimes b) := g(a)(b)$$

for $g \in \mathrm{Hom}_{kG}\big(A, \mathrm{Hom}(B, C)\big)$, $a \in A$, $b \in B$. We must verify that $\psi(f)$ and $\theta(g)$ are kG-homomorphisms. Let $x \in G$.

$$\begin{aligned}
\Big(\big(x\psi(f)\big)(a)\Big)(b) &= \Big(x\big(\psi(f)(x^{-1}a)\big)\Big)(b) = x\Big(\big(\psi(f)(x^{-1}a)\big)(x^{-1}b)\Big) \\
&= x\big(f(x^{-1}a \otimes x^{-1}b)\big) = x\Big(f\big(x^{-1}(a \otimes b)\big)\Big) \\
&= (xf)(a \otimes b) = f(a \otimes b) = \big(\psi(f)(a)\big)(b).
\end{aligned}$$

Similarly, we have $\big(x\theta(g)\big)(a \otimes b) = \theta(g)(a \otimes b)$. Moreover ψ, θ are inverses of each other, and the isomorphisms are natural.

The other natural isomorphism is induced from the natural isomorphism in Proposition 2.1.

For the statement about Ext, let $P_* \xrightarrow{\varepsilon} A$ be a projective resolution of A. Then $P_* \otimes B \xrightarrow{\varepsilon\otimes 1} A \otimes B$ is a projective resolution of $A \otimes B$. By the first part and the naturality

$$\mathrm{Hom}_{kG}(P_* \otimes B, C) \cong \mathrm{Hom}_{kG}(P_*, \mathrm{Hom}(B, C))$$

as complexes. Hence they have isomorphic homology. □

Here is a list of possible inner products.

Suppose that $\zeta \in \mathrm{Ext}^m_{kG}(M, N)$ and $\gamma \in \mathrm{Ext}^n_{kG}(L, M)$.

(5) *Outer product with composition.* Define the product of ζ and γ by the composition

$$\begin{aligned}\mathrm{Ext}^m_{kG}(M,N) \otimes \mathrm{Ext}^n_{kG}(L,M) &\cong \mathrm{Ext}^m_{kG}(k, \mathrm{Hom}(M,N)) \otimes \mathrm{Ext}^n_{kG}(k, \mathrm{Hom}(L,M)) \\ &\xrightarrow{\text{outer product}} \mathrm{Ext}^{m+n}_{kG}(k, \mathrm{Hom}(M,N) \otimes \mathrm{Hom}(L,M)) \\ &\xrightarrow{\text{composition}} \mathrm{Ext}^{m+n}_{kG}(k, \mathrm{Hom}(L,N)) \cong \mathrm{Ext}^{m+n}_{kG}(L,N).\end{aligned}$$

(6) *Composition of maps in the stable category.* Here we assume that $m > 0$, $n > 0$. Define the product of ζ and γ by the composition

$$\begin{aligned}\mathrm{Ext}^m_{kG}(M,N) \otimes \mathrm{Ext}^n_{kG}(L,M) &\cong \underline{\mathrm{Hom}}_{kG}(\Omega^m(M), N) \otimes \underline{\mathrm{Hom}}_{kG}(\Omega^n(L), M) \\ &\cong \underline{\mathrm{Hom}}_{kG}(\Omega^m(M), N) \otimes \underline{\mathrm{Hom}}_{kG}(\Omega^{m+n}(L), \Omega^m(M)) \\ &\xrightarrow{\text{composition}} \underline{\mathrm{Hom}}_{kG}(\Omega^{m+n}(L), N) \cong \mathrm{Ext}^{m+n}_{kG}(L,N).\end{aligned}$$

(7) *Composition of chain maps.* Let $P_* \xrightarrow{\varepsilon_L} L$, $Q_* \xrightarrow{\varepsilon_M} M$, and $R_* \xrightarrow{\varepsilon_N} N$ be projective resolutions. Then ζ is represented by a chain homomorphism of degree $-m$, say $\zeta_* : Q_* \to R_*$, and γ is represented by a chain homomorphism of degree $-n$, say $\gamma_* : P_* \to Q_*$. Let $\zeta\gamma$ be the class of the chain homomorphism $\zeta_* \circ \gamma_*$.

(8) *Yoneda splice.* Let ζ be represented by

$$E_\zeta : \ 0 \longrightarrow N \longrightarrow A_{m-1} \longrightarrow \cdots \longrightarrow A_0 \xrightarrow{\pi} M \longrightarrow 0,$$

and let γ be represented by

$$E_\gamma : \ 0 \longrightarrow M \xrightarrow{\iota'} B_{n-1} \longrightarrow \cdots \longrightarrow B_0 \longrightarrow L \longrightarrow 0.$$

Then let $\zeta\gamma$ be the class of the sequence

$$E_\zeta \circ E_\gamma : \ 0 \to N \to A_{m-1} \to \cdots \to A_0 \xrightarrow{\iota'\circ\pi} B_{n-1} \to \cdots \to B_0 \to L \to 0.$$

$$A_0 \searrow M \nearrow B_{n-1}$$

Theorem 6.6 *All inner products* (5), (6), (7), *and* (8) *coincide.*

Proof. We show that (5) and (7) are equivalent. The diagram we construct should give a hint as to the other equivalences.

We have $\zeta \in \mathrm{Ext}^m_{kG}(M,N)$ and $\gamma \in \mathrm{Ext}^n_{kG}(L,M)$. Let $X_* \overset{\varepsilon}{\twoheadrightarrow} k$ be a minimal projective resolution of the trivial kG-module k.

First, we consider the composition of chain maps. The rows in the next diagram are projective resolutions of L, M, and N. The maps $\hat{\zeta}$ and $\hat{\gamma}$ should be taken as representatives for ζ and γ. The composition of chain maps is determined by the composition $\hat{\zeta} \circ (1 \otimes \hat{\gamma}) : \Omega^m(k) \otimes \Omega^n(k) \otimes L \to N$.

$$\begin{array}{ccccccccccccc}
\cdots \to X_0 \otimes \Omega^m(k) \otimes \Omega^n(k) \otimes L & \to & X_{m-1} \otimes \Omega^n(k) \otimes L \to & \cdots & \to X_0 \otimes \Omega^n(k) \otimes L & \to & X_{n-1} \otimes L \to & \cdots & \to X_0 \otimes L \xrightarrow{\varepsilon \otimes 1} k \otimes L \to 0 \\
\downarrow & \searrow \ \Omega^m(k) \otimes \Omega^n(k) \otimes L \ \nearrow & \downarrow{\scriptstyle 1\otimes\hat{\gamma}} & & \downarrow{\scriptstyle 1\otimes\hat{\gamma}} & \searrow \ \Omega^n(k) \otimes L \ \nearrow & & & \\
 & & & & & \downarrow{\scriptstyle \hat{\gamma}} & & & \\
\cdots \to X_0 \otimes \Omega^m(k) \otimes M & \longrightarrow & X_{m-1} \otimes M & \to \cdots \to & X_0 \otimes M & \longrightarrow & M & \longrightarrow & 0 \\
\downarrow{\scriptstyle 1\otimes\hat{\zeta}} & \searrow \ \Omega^m(k) \otimes M \ \nearrow \ ({\scriptstyle 1\otimes\hat{\gamma}}\downarrow) & & & & & & & \\
 & \downarrow{\scriptstyle \hat{\zeta}} & & & & & & & \\
\cdots \to X_0 \otimes N & \longrightarrow & N & \longrightarrow & 0 & & & &
\end{array}$$

We read from the diagram that the composition of the chain maps is determined by the cocycle

$$\hat{\zeta} \circ (1 \otimes \hat{\gamma}) \in \mathrm{Hom}_{kG}\big(\Omega^m(k) \otimes \Omega^n(k) \otimes L, N\big).$$

Now look at the following diagram.

$$\begin{array}{ccc}
\underline{\mathrm{Hom}}_{kG}\big(\Omega^m(k) \otimes M, N\big) \otimes \underline{\mathrm{Hom}}_{kG}\big(\Omega^n(k) \otimes L, M\big) & \xrightarrow{\psi_1 \otimes \psi_2} & \underline{\mathrm{Hom}}_{kG}\big(\Omega^m(k), \mathrm{Hom}(M,N)\big) \otimes \underline{\mathrm{Hom}}_{kG}\big(\Omega^n(k), \mathrm{Hom}(L,M)\big) \\
\downarrow{\scriptstyle \varphi} & & \\
\underline{\mathrm{Hom}}_{kG}\big(\Omega^m(k) \otimes M, N\big) \otimes \underline{\mathrm{Hom}}_{kG}\big(\Omega^m(k) \otimes \Omega^n(k) \otimes L, \Omega^m(k) \otimes M\big) & & \downarrow{\scriptstyle \rho_2} \\
\downarrow{\scriptstyle \rho_1} & & \\
\underline{\mathrm{Hom}}_{kG}\big(\Omega^m(k) \otimes \Omega^n(k) \otimes L, N\big) & \xrightarrow{\psi_3} & \underline{\mathrm{Hom}}_{kG}\big(\Omega^m(k) \otimes \Omega^n(k), \mathrm{Hom}(L,N)\big)
\end{array}$$

Here φ sends $\hat{\zeta}\otimes\hat{\gamma}$ to $\hat{\zeta}\otimes(1\otimes\hat{\gamma})$, ρ_1 is composition, ρ_2 is induced by composition, and ψ_1, ψ_2, ψ_3 are the usual isomorphisms. The map $\psi_3 \circ \rho_1 \circ \varphi$ is equivalent to the composition of the chain maps, i.e., it corresponds to (7). On the other hand $\rho_2 \circ (\psi_1 \otimes \psi_2)$ is the outer product with composition, i.e., it corresponds to (5).

We show that the diagram commutes. For $f \in \underline{\mathrm{Hom}}_{kG}(\Omega^m(k) \otimes M, N)$, $g \in \underline{\mathrm{Hom}}_{kG}(\Omega^n(k) \otimes L, M)$, $a \in \Omega^m(k)$, $b \in \Omega^n(k)$, $l \in L$ we compute

$$\begin{aligned}
&\big(\psi_3 \circ \rho_1 \circ \varphi(f \otimes g)\big)(a \otimes b)(l) = \big(\psi_3(f \circ (1 \otimes g))\big)(a \otimes b)(l) \\
&\quad = f \circ (1 \otimes g)(a \otimes b \otimes l) = f\big(a \otimes g(b \otimes l)\big) \qquad \text{and} \\
&\big(\rho_2 \circ (\psi_1 \otimes \psi_2)(f \otimes g)\big)(a \otimes b)(l) = \big(\rho_2 \circ (\psi_1(f) \otimes \psi_2(g))\big)(a \otimes b)(l) \\
&\quad = \big(\psi_1(f)(a) \circ \psi_2(g)(b)\big)(l) = \big(\psi_1(f)(a)\big)\big(g(b \otimes l)\big) = f\big(a \otimes g(b \otimes l)\big). \quad \square
\end{aligned}$$

Corollary 6.7 *The outer and inner products are associative.*

Proof. Look at (4) for the outer products and at (7) or (8) for the inner ones. □

Proposition 6.8 *The outer products are graded commutative. So, for example, if $\zeta \in \mathrm{Ext}^m_{kG}(M, N)$ and $\gamma \in \mathrm{Ext}^n_{kG}(M', N')$ are represented by extensions E_ζ and E_γ, respectively, then*

$$\mathrm{class}\big((E_\zeta \otimes N') \circ (M \otimes E_\gamma)\big) = (-1)^{mn}\mathrm{class}\big((N \otimes E_\gamma) \circ (E_\zeta \otimes M')\big).$$

Proof. Let $P_* \overset{\varepsilon}{\twoheadrightarrow} M$ and $P'_* \overset{\varepsilon'}{\twoheadrightarrow} M'$ be projective resolutions. Define

$$\sigma_* : (P_* \otimes P'_* \xrightarrow{\varepsilon\otimes\varepsilon'} M \otimes M') \longrightarrow (P'_* \otimes P_* \xrightarrow{\varepsilon'\otimes\varepsilon} M' \otimes M)$$

on homogeneous elements $x \in P_m$, $y \in P'_n$ by

$$\sigma_{m+n}(x \otimes y) := (-1)^{mn} y \otimes x.$$

Clearly σ_* lifts the map $M \otimes M' \xrightarrow{\text{flip}} M' \otimes M$ given by $m \otimes m' \mapsto m' \otimes m$. We need to check that σ_* is a chain homomorphism.

$$\begin{aligned}
\sigma \circ \partial(x \otimes y) &= \sigma\big(\partial x \otimes y + (-1)^m x \otimes \partial y\big) \\
&= (-1)^{(m-1)n} y \otimes \partial x + (-1)^m (-1)^{m(n-1)} \partial y \otimes x \\
&= (-1)^{mn}\big(\partial y \otimes x + (-1)^n y \otimes \partial x\big) = (-1)^{mn}\partial(y \otimes x) = \partial \circ \sigma(x \otimes y).
\end{aligned}$$

If $\hat{\zeta} : P_m \to N$ represents ζ and $\hat{\gamma} : P'_n \to N'$ represents γ, then the sequence $(N \otimes E_\gamma) \circ (E_\zeta \otimes M')$ is associated to the cocycle given by the composition

$$P_* \otimes P'_* \xrightarrow{\sigma} P'_* \otimes P_* \xrightarrow{\hat{\gamma}\otimes\hat{\zeta}} N' \otimes N \xrightarrow{\text{flip}} N \otimes N',$$

which is $(-1)^{mn}\hat{\zeta}\otimes\hat{\gamma}$ if we identify $M' \otimes M$ with $M \otimes M'$ and $N' \otimes N$ with $N \otimes N'$ by the flip maps. □

For $M = M' = N = N' = k$ we get the next corollary.

Corollary 6.9 $H^*(G,k) = \mathrm{Ext}^*_{kG}(k,k)$ *is a graded commutative ring.*

Remark The ring $\mathrm{Ext}^*_{kG}(M,M)$ is in general not graded commutative. In fact, $\mathrm{Ext}^0_{kG}(M,M) = \mathrm{Hom}_{kG}(M,M)$ is in general not commutative.

Definition For $\zeta \in H^n(G,k) - \{0\}$, $n \geqslant 1$, with $\hat{\zeta} : \Omega^n(k) \to k$ representing ζ let $L_\zeta := \ker \hat{\zeta}$, so that we have a short exact sequence

$$0 \to L_\zeta \hookrightarrow \Omega^n(k) \xrightarrow{\hat{\zeta}} k \to 0.$$

Lemma 6.10 L_ζ *is well-defined up to isomorphism.*

Proof. This follows from Proposition 5.5. In fact, the module L_ζ is the third object in the triangle for the morphism $\hat{\zeta}$. Hence the isomorphism class of L_ζ in ${}_{kG}\mathfrak{stmod}$ is determined by ζ. Since $\Omega^n(k)$ has no nonzero projective summands, L_ζ is in fact defined up to isomorphism in ${}_{kG}\mathfrak{mod}$ by ζ. □

Theorem 5.1 suggests the next definition, which we will employ in the subsequent theorem.

Definition For M, N in ${}_{kG}\mathfrak{mod}$ and $n \in \mathbb{Z}$ let

$$\widehat{\mathrm{Ext}}^n_{kG}(M,N) := \underline{\mathrm{Hom}}_{kG}\big(\Omega^n(M), N\big).$$

The functor $\widehat{\mathrm{Ext}}^n_{kG}$ is called the nth Tate cohomology. The functor $\widehat{\mathrm{Ext}}^n_{kG}$ can also be obtained as the nth cohomology of the complex $\mathrm{Hom}_{kG}(P_*, N)$ where

$$P_* : \quad \cdots \longrightarrow P_1 \longrightarrow P_0 \xrightarrow{\partial_0} P_{-1} \longrightarrow P_{-2} \longrightarrow \cdots$$

is a complete projective resolution of M, in that $\partial_0(P_0) \cong M$. Of course, $\widehat{\mathrm{Ext}}^n_{kG} \cong \mathrm{Ext}^n_{kG}$ if $n > 0$.

Theorem 6.11 ([C2]) *Let* $\zeta \in H^n(G,k) - \{0\}$, $n \geqslant 1$. *If* n *is even and* $p > 2$ *(recall that* $p = \operatorname{char} k$*), then* ζ *annihilates the cohomology of* L_ζ. *This means that* ζ *annihilates* $\widehat{\mathrm{Ext}}^*_{kG}(M, L_\zeta)$ *and* $\widehat{\mathrm{Ext}}^*_{kG}(L_\zeta, M)$ *for any* M *in* ${}_{kG}\mathfrak{mod}$.

Proof. We may assume that $p \,\big|\, |G|$, as there is nothing to prove otherwise.

It suffices to show that ζ annihilates $\mathrm{id}_{L_\zeta} \in \underline{\mathrm{Hom}}_{kG}(L_\zeta, L_\zeta) = \widehat{\mathrm{Ext}}^0_{kG}(L_\zeta, L_\zeta)$. This is because if $\gamma \in \widehat{\mathrm{Ext}}^*_{kG}(M, L_\zeta)$, then $\zeta\gamma = \zeta(\mathrm{id}_{L_\zeta} \circ \gamma) = (\zeta \cdot \mathrm{id}_{L_\zeta})\gamma = 0$ and similarly for $\widehat{\mathrm{Ext}}^*_{kG}(L_\zeta, M)$.

We have

$$\zeta \in H^n(G,k) = \operatorname{Ext}^n_{kG}(k,k) \cong \underline{\operatorname{Hom}}_{kG}\big(\Omega^n(k),k\big) \cong \operatorname{Ext}^1_{kG}\big(\Omega^{n-1}(k),k\big).$$

As an element of $\operatorname{Ext}^1_{kG}\big(\Omega^{n-1}(k),k\big)$, ζ is represented by the following extension.

$$E_\zeta : \quad 0 \to k \to \Omega^{-1}(L_\zeta) \to \Omega^{n-1}(k) \to 0. \tag{6.5}$$

To see this, let $P_* \xrightarrow{\varepsilon} k$ be a minimal projective resolution of the trivial module k. We have the following diagram (it is an exercise to show that there is no projective summand in the middle term of the bottom row):

$$\begin{array}{ccccccccc}
 & & 0 & & 0 & & & & \\
 & & \downarrow & & \downarrow & & & & \\
 & & L_\zeta & = & L_\zeta & & & & \\
 & & \downarrow & & \downarrow & & & & \\
0 & \longrightarrow & \Omega^n(k) & \longrightarrow & P_{n-1} & \longrightarrow & P_{n-2} & \to \cdots \to P_0 \to & k \to 0 \\
 & & \hat{\zeta}\downarrow & \text{pushout} & \downarrow & \searrow & \uparrow & & \\
0 & \longrightarrow & k & \longrightarrow & \Omega^{-1}(L_\zeta) & \longrightarrow & \Omega^{n-1}(k) & \longrightarrow & 0 \\
 & & \downarrow & & \downarrow & & & & \\
 & & 0 & & 0 & & & &
\end{array}$$

So $\zeta \cdot \mathrm{id}_{L_\zeta}$ is represented by the extension

$$E_\zeta \otimes L_\zeta : \quad 0 \to L_\zeta \to \Omega^{-1}(L_\zeta) \otimes L_\zeta \to \underset{\substack{\| \wr \\ \Omega^{n-1}(L_\zeta) \oplus (\text{proj})}}{\Omega^{n-1}(k) \otimes L_\zeta} \to 0. \tag{6.6}$$

Hence $\zeta \cdot \mathrm{id}_{L_\zeta} = 0$ if the short exact sequence (6.6) splits. This is the case if

$$\Omega^{-1}(L_\zeta) \otimes L_\zeta \cong L_\zeta \oplus \Omega^{n-1}(L_\zeta) \oplus (\text{proj}). \tag{6.7}$$

Let's express this criterion for when a short exact sequence splits in a lemma.

Lemma 6.12 *A short exact sequence*

$$E : \quad 0 \longrightarrow A \xrightarrow{\ \alpha\ } B \xrightarrow{\ \beta\ } C \longrightarrow 0$$

in ${}_{kG}\mathfrak{mod}$ *splits if and only if* $B \cong A \oplus C$.

Proof. If E splits, then clearly $B \cong A \oplus C$.

Suppose that $B \cong A \oplus C$. From the long exact sequence

$$0 \longrightarrow \operatorname{Hom}_{kG}(C, A) \xrightarrow{\alpha_*} \operatorname{Hom}_{kG}(C, B) \xrightarrow{\beta_*} \operatorname{Hom}_{kG}(C, C) \xrightarrow{\delta} \operatorname{Ext}^1_{kG}(C, A) \longrightarrow \cdots$$

we get

$$\begin{aligned} \dim \operatorname{im} \beta_* &= \dim \operatorname{Hom}_{kG}(C, B) - \dim \ker \beta_* \\ &= \dim \operatorname{Hom}_{kG}(C, A \oplus C) - \dim \operatorname{im} \alpha_* = \dim \operatorname{Hom}_{kG}(C, C). \end{aligned}$$

So β_* is surjective, and any preimage of $\mathrm{id}_C \in \operatorname{Hom}_{kG}(C, C)$ splits β. □

An alternative approach to the previous proof begins by first noting that we have $\operatorname{class}(E) = \delta(\mathrm{id}_C) \in \operatorname{Ext}^1_{kG}(C, A)$. We must show that $\delta(\mathrm{id}_C) = 0$. But this is obvious because $\delta = 0$.

Now we continue with the proof of Theorem 6.11. It remains to be seen that (6.7) holds, or that

$$L_\zeta \otimes L_\zeta \cong \Omega(L_\zeta) \oplus \Omega^n(L_\zeta) \oplus (\mathrm{proj}). \tag{6.8}$$

To show that (6.8) implies (6.7) note that $\Omega^0(L_\zeta) = L_\zeta$.

Consider the complex

$$C_* : \quad 0 \longrightarrow \Omega^n(k) \xrightarrow{\hat{\zeta}} k \longrightarrow 0 \qquad \text{with } k \text{ in degree } 0,$$

so that $H_*(C_*) = H_1(C_*) = L_\zeta$. The square $C_* \otimes C_*$ of C_*,

$$0 \longrightarrow \Omega^n(k) \otimes \Omega^n(k) \xrightarrow{\binom{\hat{\zeta}\otimes 1}{-1\otimes\hat{\zeta}}} \begin{matrix} k \otimes \Omega^n(k) \\ \oplus \\ \Omega^n(k) \otimes k \end{matrix} \xrightarrow{(1\otimes\hat{\zeta}\ \ \hat{\zeta}\otimes 1)} k \otimes k \longrightarrow 0,$$

has homology $H_*(C_* \otimes C_*) \cong H_*(C_*) \otimes H_*(C_*) = L_\zeta \otimes L_\zeta$ in degree 2.

Let $\sigma_* : C_* \otimes C_* \longrightarrow C_* \otimes C_*$ be the chain map with $\sigma_{m+n}(x \otimes y) = (-1)^{mn} y \otimes x$ for $x \in C_m$ and $y \in C_n$. So $C_* \otimes C_*$ is the direct sum of the (± 1)-eigenspaces $D_*^+ = S^2 C_*$ and $D_*^- = \Lambda^2 C_*$ of σ_*.

$$D_n^+ = \bigoplus_{i+j=n} \left\{ x \otimes y + (-1)^{ij} y \otimes x \;\middle|\; x \in C_i, y \in C_j \right\}$$

$$D_n^- = \bigoplus_{i+j=n} \left\{ x \otimes y - (-1)^{ij} y \otimes x \;\middle|\; x \in C_i, y \in C_j \right\}$$

So clearly $D_0^+ = k \otimes k$, $D_0^- = 0$, $D_1^+ \cong \Omega^n(k) \cong D_1^-$. We need to look at $D_2^+ \oplus D_2^- = (C_* \otimes C_*)_2 = \Omega^n(k) \otimes \Omega^n(k) \cong \Omega^{2n}(k) \oplus (\mathrm{proj})$. The exact sequence

$$0 \longrightarrow \Omega^n(k) \longrightarrow P_{n-1} \longrightarrow P_{n-2} \longrightarrow \cdots \longrightarrow P_0 \longrightarrow k \longrightarrow 0$$

shows that $\dim \Omega^n(k) = \Big(\sum_{i=0}^{n-1}(-1)^i \dim P_{n-1-i}\Big)+(-1)^n$. Since projective modules have dimensions divisible by p and n is even, we get $\dim \Omega^n(k) \equiv 1 \,(\mathrm{mod}\, p)$. Hence

$$\dim D_2^+ = \frac{\dim \Omega^n(k)\big(\dim \Omega^n(k) - 1\big)}{2}$$

is divisible by p and $\dim D_2^- \equiv 1 \,(\mathrm{mod}\, p)$. Since $\Omega^{2n}(k)$ is indecomposable, we get that $D_2^- \cong \Omega^{2n}(k) \oplus (\mathrm{proj})$ and $D_2^+ = (\mathrm{proj})$.

The two subcomplexes D_*^+, D_*^- of $C_* \otimes C_*$ are isomorphic to

$$D_*^+ : \; 0 \longrightarrow (\mathrm{proj}) \xrightarrow{\alpha} \Omega^n(k) \xrightarrow{\tilde{\zeta}} k \longrightarrow 0$$

$$D_*^- : \; 0 \longrightarrow \Omega^{2n}(k) \oplus (\mathrm{proj}) \longrightarrow \Omega^n(k) \longrightarrow 0 \longrightarrow 0,$$

whose homologies are concentrated in degree 2 and with $\tilde{\zeta}$ equivalent to $\hat{\zeta}$. In particular, we have $\operatorname{im} \alpha = \ker \tilde{\zeta} \cong L_\zeta$, so that $H_2(D_*^+) \cong \Omega(L_\zeta) \oplus (\mathrm{proj})$. Also $H_2(D_*^-) \cong \Omega^n(L_\zeta) \oplus (\mathrm{proj})$. Putting things together, we finally have that

$$\begin{aligned} L_\zeta \otimes L_\zeta &\cong H_2(C_* \otimes C_*) = H_2(D_*^+ \oplus D_*^-) \\ &\cong H_2(D_*^+) \oplus H_2(D_*^-) \cong \Omega(L_\zeta) \oplus \Omega^n(L_\zeta) \oplus (\mathrm{proj}), \end{aligned}$$

which is (6.8). □

The proof of Theorem 6.11 can be adapted to show also that if $p > 2$ and n is odd, then no nonzero element $\zeta \in H^n(G,k)$ has the property that ζ annihilates the cohomology of L_ζ. For $p = 2$, the situation is far more complicated, and in general the question of when ζ annihilates the cohomology of L_ζ is an open one. However, we can prove the following proposition.

Proposition 6.13 *Suppose that $\zeta \in H^n(G,k)$, $n \geqslant 1$. Then ζ^2 annihilates the cohomology of L_ζ.*

Proof. Assume that $\zeta \neq 0$, as the result is obvious otherwise. Then we have an exact sequence

$$0 \longrightarrow L_\zeta \xrightarrow{\sigma} \Omega^n(k) \xrightarrow{\hat{\zeta}} k \longrightarrow 0$$

where $\hat{\zeta}$ is a cocycle representing ζ. Applying $\widehat{\mathrm{Ext}}^*_{kG}(L_\zeta, \;\;)$ we obtain a corresponding long exact sequence and a commutative diagram

$$\begin{array}{ccccccc} \cdots \xrightarrow{\zeta_*} & \widehat{\mathrm{Ext}}^{m-1}_{kG}(L_\zeta,k) & \longrightarrow & \widehat{\mathrm{Ext}}^{m}_{kG}(L_\zeta,L_\zeta) & \xrightarrow{\sigma_*} & \widehat{\mathrm{Ext}}^{m}_{kG}(L_\zeta,\Omega^n(k)) & \xrightarrow{\zeta_*} \cdots \\ & \downarrow{\scriptstyle\zeta} & & \downarrow{\scriptstyle\zeta} & & \downarrow{\scriptstyle\zeta} & \\ \cdots \xrightarrow{\zeta_*} & \widehat{\mathrm{Ext}}^{m+n-1}_{kG}(L_\zeta,k) & \xrightarrow{\mu_*} & \widehat{\mathrm{Ext}}^{m+n}_{kG}(L_\zeta,L_\zeta) & \xrightarrow{\sigma_*} & \widehat{\mathrm{Ext}}^{m+n}_{kG}(L_\zeta,\Omega^n(k)) & \xrightarrow{\zeta_*} \cdots \end{array}$$

where the vertical maps are all multiplication by ζ. Notice that $\widehat{\mathrm{Ext}}^m_{kG}(L_\zeta, \Omega^n(k)) \cong \widehat{\mathrm{Ext}}^{m-n}_{kG}(L_\zeta, k)$, and the connecting homomorphisms are likewise multiplications by ζ. If $\gamma \in \widehat{\mathrm{Ext}}^m_{kG}(L_\zeta, L_\zeta)$, then $\sigma_*(\gamma)$ is in the kernel of multiplication by ζ. So $\tilde{\zeta} \circ \sigma_*(\gamma) = 0$. It follows that $\tilde{\zeta} \circ \gamma = \mu_*(\beta)$ for some $\beta \in \widehat{\mathrm{Ext}}^{m+n-1}_{kG}(L_\zeta, k)$. But then $\zeta^2\gamma$ is represented by $\zeta(\tilde{\zeta} \circ \gamma) = \mu_* \circ \zeta_*(\tilde{\beta}) = 0$, where $\tilde{\beta} \in \widehat{\mathrm{Ext}}^{m+2n-1}_{kG}(L_\zeta, \Omega^n(k))$ corresponds to β. □

7 Examples and diagrams

Before proceeding further we pause here to consider some examples. Except in a few special cases, cohomology rings are very difficult to calculate. Fortunately, one of the special cases is that of an elementary abelian p-group. These groups play a major role in the cohomology theory, and so we want to look at their cohomology rings here. Also we consider certain group algebras whose projective modules can be represented by diagrams. For these we can compute products in cohomology as compositions of chain maps. In all of this remember that the base field k has characteristic p.

As our first example we take a cyclic group of order $p^n > 1$.

Proposition 7.1 *The group $G = \langle g \mid g^{p^n} = 1 \rangle$, $n > 0$, has a minimal projective resolution of the form*

$$\begin{array}{ccccccccccc} \cdots \xrightarrow{(g-1)^{p^n-1}} & kG & \xrightarrow{g-1} & kG & \xrightarrow{(g-1)^{p^n-1}} & kG & \xrightarrow{g-1} & kG & \xrightarrow{\varepsilon} & k & \longrightarrow 0 \\ & \| & & \| & & \| & & \| & & & \\ & X_3 & & X_2 & & X_1 & & X_0 & & & \end{array}$$

Proof. It's a complex because $(g-1)^{p^n} = g^{p^n} - 1 = 1 - 1 = 0$. The exactness follows from the fact that $1, g-1, (g-1)^2, \ldots, (g-1)^{p^n-1}$ is a k-basis for kG. In fact,

$$\mathrm{im}(g-1) = \bigoplus_{i=1}^{p^n-1} k(g-1)^i = \ker(g-1)^{p^n-1}$$

and

$$\mathrm{im}(g-1)^{p^n-1} = k(g-1)^{p^n-1} = \ker(g-1).$$

Exactness at the rightmost copy of kG and the minimality are also clear. □

Remark Note that $(g-1)^{p^n-1} = \sum_{i=0}^{p^n-1} g^i = \widetilde{G}$. If instead of k we took $\mathbb{Z}$ as our base ring, then we would have a projective resolution

$$\cdots \xrightarrow{\widetilde{G}} \mathbb{Z}G \xrightarrow{g-1} \mathbb{Z}G \xrightarrow{\widetilde{G}} \mathbb{Z}G \xrightarrow{g-1} \mathbb{Z}G \xrightarrow{\varepsilon} \mathbb{Z} \to 0$$

of $G = \langle\, g \mid g^N = 1 \,\rangle$ where $\widetilde{G} = \sum_{i=0}^{N-1} g^i$.

Again let $G = \langle\, g \mid g^{p^n} = 1 \,\rangle$. We can write kG as a truncated polynomial algebra, namely, letting $Y \mapsto g-1$ we have an isomorphism $k[Y]/(Y^{p^n}) \cong kG$.

This observation leads to the notion of a diagram for a module. In a diagram we let a vertex represent an element of a k-basis for the module. An edge (or arrow) represents multiplication by some specified element in $\operatorname{rad} kG - \operatorname{rad}^2 kG$. Of course, the number of vertices must be the dimension of the module. The labels on the arrows should denote elements of a basis of $\operatorname{rad} kG/\operatorname{rad}^2 kG$. In the case of the cyclic group G the basis consists of the single element Y.

Here is the diagram for $kG \cong k[Y]/(Y^{p^n})$:

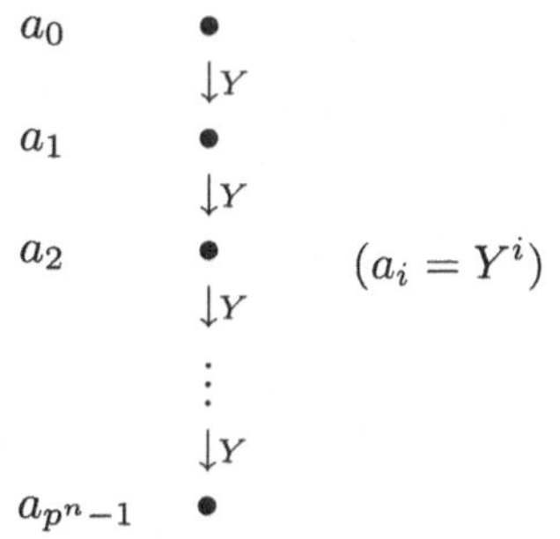

Now let's look again at the minimal projective resolution in Proposition 7.1. In diagrammatic form it looks like the following.

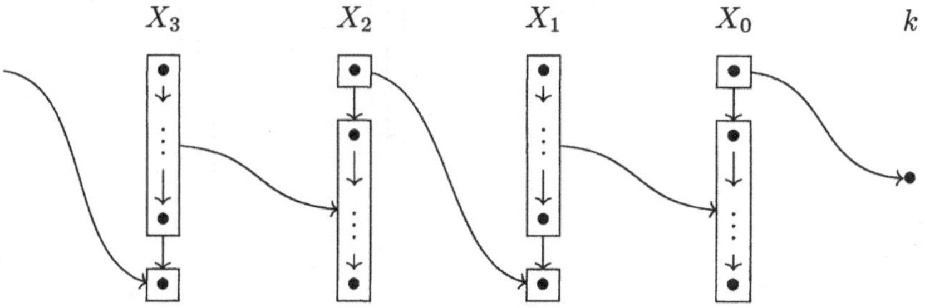

Lemma 7.2 *Let* $G = \langle\, g \mid g^{p^n} = 1 \,\rangle$, $n > 0$. *Then*

$$H^*(G,k) = \bigoplus_{i=0}^{\infty} k\gamma_i$$

where $0 \neq \gamma_i \in H^i(G,k)$.

Proof. Let $X_* \xrightarrow{\varepsilon} k$ be a minimal projective resolution as above, that is, $X_i = kGa_0 = \bigoplus_{j=0}^{p^n-1} ka_j$. To give a kG-homomorphism $X_i \longrightarrow k$ we must simply specify the image of a_0. Let $\gamma_i \in H^i(G, k)$ be represented by $\gamma_i : X_i \longrightarrow k$, $a_0 \mapsto 1$. Surely this defines a cocycle, and there are no nonzero coboundaries. □

To obtain the ring structure of $H^*(G, k)$, we represent the cohomology elements by chain maps, which we then can compose. Here is the result.

Theorem 7.3 *Let $G = \langle g \mid g^{p^n} = 1 \rangle$, $n > 0$. Then*

$$H^*(G,k) \cong \begin{cases} k[\gamma_1] & \text{if } p^n = 2, \\ k[\gamma_1, \gamma_2]/(\gamma_1^2) & \text{if } p^n > 2. \end{cases}$$

with $\deg \gamma_i = i$.

Proof. The chain map for γ_2 is given by $\hat{\gamma}_2 : X_{i+2} \longrightarrow X_i$, $a_0 \mapsto a_0$ for $i \geqslant 0$. Hence $\gamma_i \gamma_2 = \gamma_{i+2}$ for $i \geqslant 0$.

If $p^n = 2$, then we get $\gamma_i \gamma_1 = \gamma_{i+1}$ for $i \geqslant 0$, by a similar chain map in degree 1.

It remains to be seen that $\gamma_1^2 = 0$ if $p^n > 2$. For this we construct the chain map which covers γ_1. In diagrammatic form it looks like this:

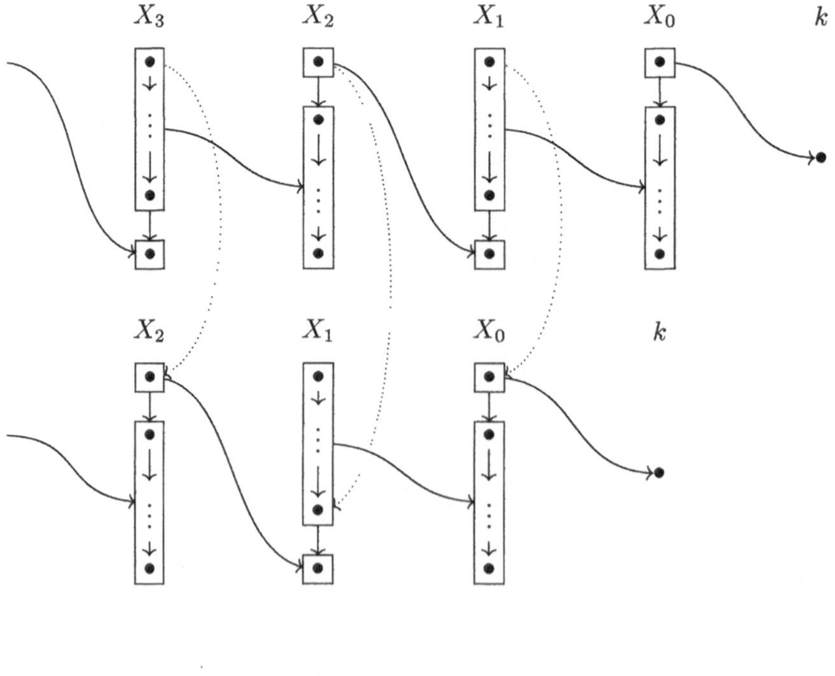

Therefore

$$\begin{array}{ccccc} X_2 & \xrightarrow{\hat{\gamma}_1} & X_1 & \xrightarrow{\gamma_1} & k \\ a_0 & \longmapsto & a_{p^n-2} & \longmapsto & 0 \qquad \text{if } p^n > 2. \end{array}$$

So $\gamma_1^2 = 0$ in this case. □

Now suppose that $G = \langle y_1, \ldots, y_n \rangle \cong (\mathbb{Z}/p)^n$ is an elementary abelian p-group. For each $i \in \{1, \ldots, n\}$ let $H_i := \langle y_i \rangle \subseteq G$, so that $G = H_1 \times \cdots \times H_n$.

Proposition 7.4 *Let A, B be groups. Then $k(A \times B) \cong kA \otimes kB$.*

Proof. The isomorphism is given by $(a, b) \mapsto a \otimes b$ for $a \in A$, $b \in B$. □

Note that for G elementary abelian as above, we may regard

$$kH_i \cong k \otimes \cdots \otimes k \otimes kH_i \otimes k \otimes \cdots \otimes k$$

as a kG-module in which y_i acts in the usual way, and y_j acts trivially for $j \neq i$.

Proposition 7.5 *Let $X_*^{(i)} \xrightarrow{\varepsilon_i} k$ be a minimal projective resolution of the trivial kH_i-module k. Then $X_* = X_*^{(1)} \otimes \cdots \otimes X_*^{(n)} \xrightarrow{\varepsilon_1 \otimes \cdots \otimes \varepsilon_n} k \otimes \cdots \otimes k$ is a minimal projective resolution of the trivial kG-module k.*

Proof. The Künneth Theorem shows that X_* has the right homology. So we need only note that for each m

$$X_m = \bigoplus_{j_1 + \cdots + j_n = m} X_{j_1}^{(1)} \otimes \cdots \otimes X_{j_n}^{(n)} \cong \bigoplus_{j_1 + \cdots + j_n = m} kG$$

is a projective (free) kG-module.

The minimality follows from the fact that $\partial(X_m) \subseteq \operatorname{rad}(X_{m-1})$ for all $m > 0$. That is, $\operatorname{rad}(X_{m-1})$ is the intersection of the kernels of all of the homomorphisms $\gamma_{j_1}^{(1)} \otimes \cdots \otimes \gamma_{j_n}^{(n)} : X_{m-1} \to k$ with $j_1 + \cdots + j_n = m - 1$. But it is easy to see that $\left(\gamma_{j_1}^{(1)} \otimes \cdots \otimes \gamma_{j_n}^{(n)}\right) \circ \partial = 0$. If X_* were not minimal, then we would have that $X_* = P_* \oplus Q_*$ for P_* a minimal projective resolution and Q_* an exact complex of projective (free) modules. If Q_* were not zero, then it would have to be the case that $\partial(Q_m) \subseteq \operatorname{rad}(Q_{m-1})$ for m the least integer with $Q_{m-1} \neq 0$. □

Theorem 7.6 *Let $G = (\mathbb{Z}/p)^n$. Then*

$$H^*(G, k) \cong \begin{cases} k[\eta_1, \ldots, \eta_n] & \text{if } p = 2, \\ k[\zeta_1, \ldots, \zeta_n] \otimes \Lambda(\eta_1, \ldots, \eta_n) & \text{if } p > 2, \end{cases}$$

where $\eta_1, \ldots, \eta_n$ *are in degree* 1 *and* $\zeta_1, \ldots, \zeta_n$ *are in degree* 2. *Here*

$$\Lambda(\eta_1, \ldots, \eta_n) = k\langle \eta_1, \ldots, \eta_n \rangle \;/\; (\eta_i^2, \eta_i\eta_j + \eta_j\eta_i \mid i, j = 1, \ldots, n)$$

is an exterior algebra.

Proof. We use the minimal projective resolution $X_* \twoheadrightarrow k$ with

$$X_m = \bigoplus_{j_1+\cdots+j_n=m} X_{j_1}^{(1)} \otimes \cdots \otimes X_{j_n}^{(n)}$$

as in Proposition 7.5. So $\mathrm{Hom}_{kG}(X_m, k)$ is generated as a k-vector space by a set of elements indexed by the collection of all n-tuples $(j_1, \ldots, j_n)$ of non-negative integers with $\sum_{i=1}^{n} j_i = m$. These generators correspond to the products $\gamma_{j_1}^{(1)} \otimes \cdots \otimes \gamma_{j_n}^{(n)}$ where $\gamma_{j_i}^{(i)} \in H^{j_i}(H_i, k)$ and with $\gamma_{j_i}^{(i)}$ represented by the cocycle $X_{j_i}^{(i)} \to k$, $a_0 \mapsto 1$.

Each $\gamma_{j_i}^{(i)}$ is also represented by a chain map $X_*^{(i)} \to X_*^{(i)}$ of degree $-j_i$. Hence $\gamma_{j_1}^{(1)} \otimes \cdots \otimes \gamma_{j_n}^{(n)}$ is represented by the product of the chain maps. For the notation we want

$$\eta_i : X_1 \xrightarrow{\text{proj}} X_0^{(1)} \otimes \cdots \otimes X_0^{(i-1)} \otimes X_1^{(i)} \otimes X_0^{(i+1)} \otimes \cdots \otimes X_0^{(n)} \xrightarrow{\varepsilon_1 \otimes \cdots \otimes \varepsilon_{i-1} \otimes \gamma_1^{(i)} \otimes \varepsilon_{i+1} \otimes \cdots \otimes \varepsilon_n} k$$

and

$$\zeta_i : X_2 \xrightarrow{\text{proj}} X_0^{(1)} \otimes \cdots \otimes X_0^{(i-1)} \otimes X_2^{(i)} \otimes X_0^{(i+1)} \otimes \cdots \otimes X_0^{(n)} \xrightarrow{\varepsilon_1 \otimes \cdots \otimes \varepsilon_{i-1} \otimes \gamma_2^{(i)} \otimes \varepsilon_{i+1} \otimes \cdots \otimes \varepsilon_n} k$$

So the presentation of $H^*(G, k)$ follows from Theorem 7.3 together with the commutativity relations. □

Corollary 7.7 *Let* $G = (\mathbb{Z}/p)^n$. *Then* $H^*(G, k)/\operatorname{rad} H^*(G, k)$ *is a polynomial ring in* n *variables.*

Proof. If $p = 2$ then $\operatorname{rad} H^*(G, k) = 0$ and $H^*(G, k)$ is a polynomial ring in n variables. If $p > 2$ then, in the notation of Theorem 7.6, $(\eta_1, \ldots, \eta_n)$ is a nilpotent ideal, in fact, $(\eta_1, \ldots, \eta_n) = \operatorname{rad} H^*(G, k)$. Hence $H^*(G, k)/\operatorname{rad} H^*(G, k) \cong k[\zeta_1, \ldots, \zeta_n]$ is a polynomial ring in n variables in this case, too. □

We can draw diagrams for some of the modules. For example, suppose that we have $p = 2$ and $G = (\mathbb{Z}/2)^2 = \langle y_1, y_2 \rangle$. The diagram for the free module $kG \cong k[Y_1, Y_2]/(Y_1^2, Y_2^2)$ looks like this:

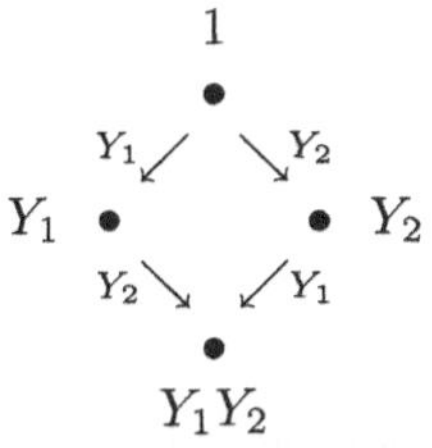

or simply

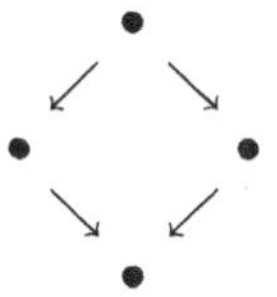

(Y_1 is $y_1 + 1$, and Y_2 is $y_2 + 1$.) Here is an example for another kG-module. It is the diagram for the module $\Omega^{-3}(k)$.

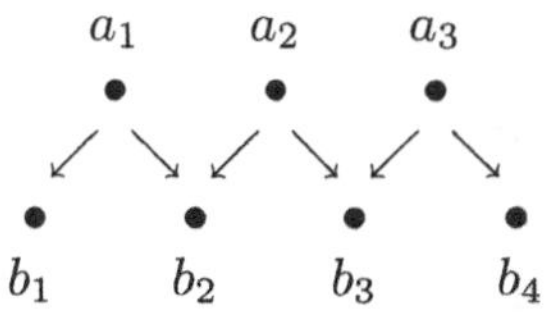

$$Y_1 a_i = b_i\,, \quad Y_2 a_i = b_{i+1} \qquad \text{for } i = 1, 2, 3,$$
$$Y_1 b_j = 0\,, \quad Y_2 b_j = 0 \qquad \text{for } j = 1, \ldots, 4.$$

So a minimal projective resolution of the trivial module looks like the following.

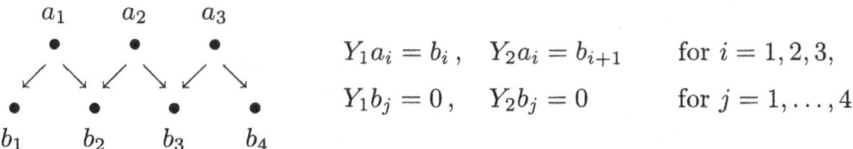

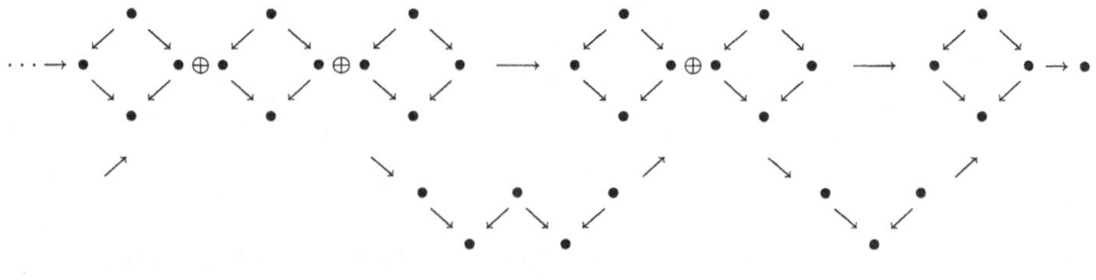

Next let's look at the dihedral group of order 8,

$$G = D_8 = \langle x, y \mid x^2 = y^2 = (xy)^4 = 1 \rangle$$

for $p = 2$. Let $A := x + 1$, $B := y + 1$. Then one computes that

$$kG = k\langle A, B \rangle / (A^2, B^2, ABAB + BABA)$$

and

$$\operatorname{rad} kG = (A, B).$$

Here $k\langle A, B \rangle$ is the polynomial ring over k in the noncommuting variables A and B.

Again, our aim is to determine the cohomology ring by looking at diagrams. The diagram for the free module kG looks like this:

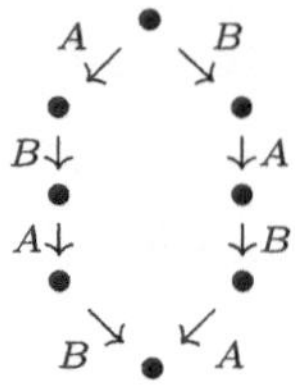

So a minimal projective resolution of the trivial module k is a splice of the sequences

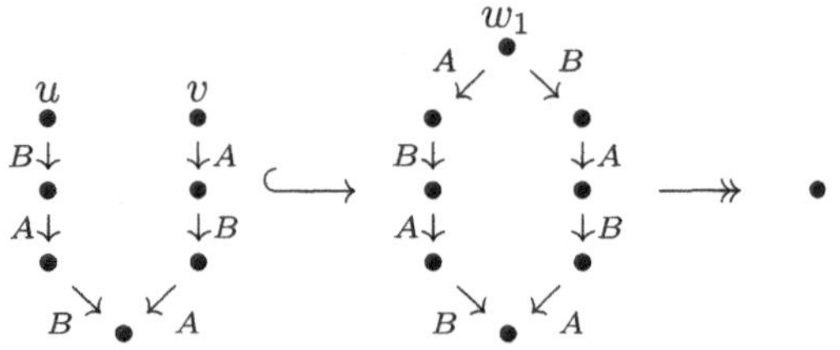

(where $u \mapsto Aw_1$ and $v \mapsto Bw_1$)

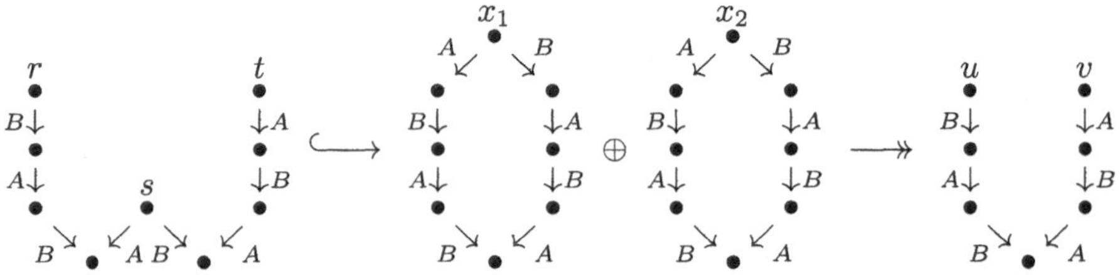

(where $x_1 \mapsto u$, $x_2 \mapsto v$, $r \mapsto Ax_1$, $s \mapsto BABx_1 + ABAx_2$, and $t \mapsto Bx_2$)

etc.

The diagram on the next page, where for simplicity the labels A and B on the arrows have been omitted, shows the chain map which represents the cohomology element a_1, given by

$$\begin{array}{rcl} a_1 : P_1 & \longrightarrow & k \\ x_1 & \longmapsto & 1 \\ x_2 & \longmapsto & 0 \end{array}$$

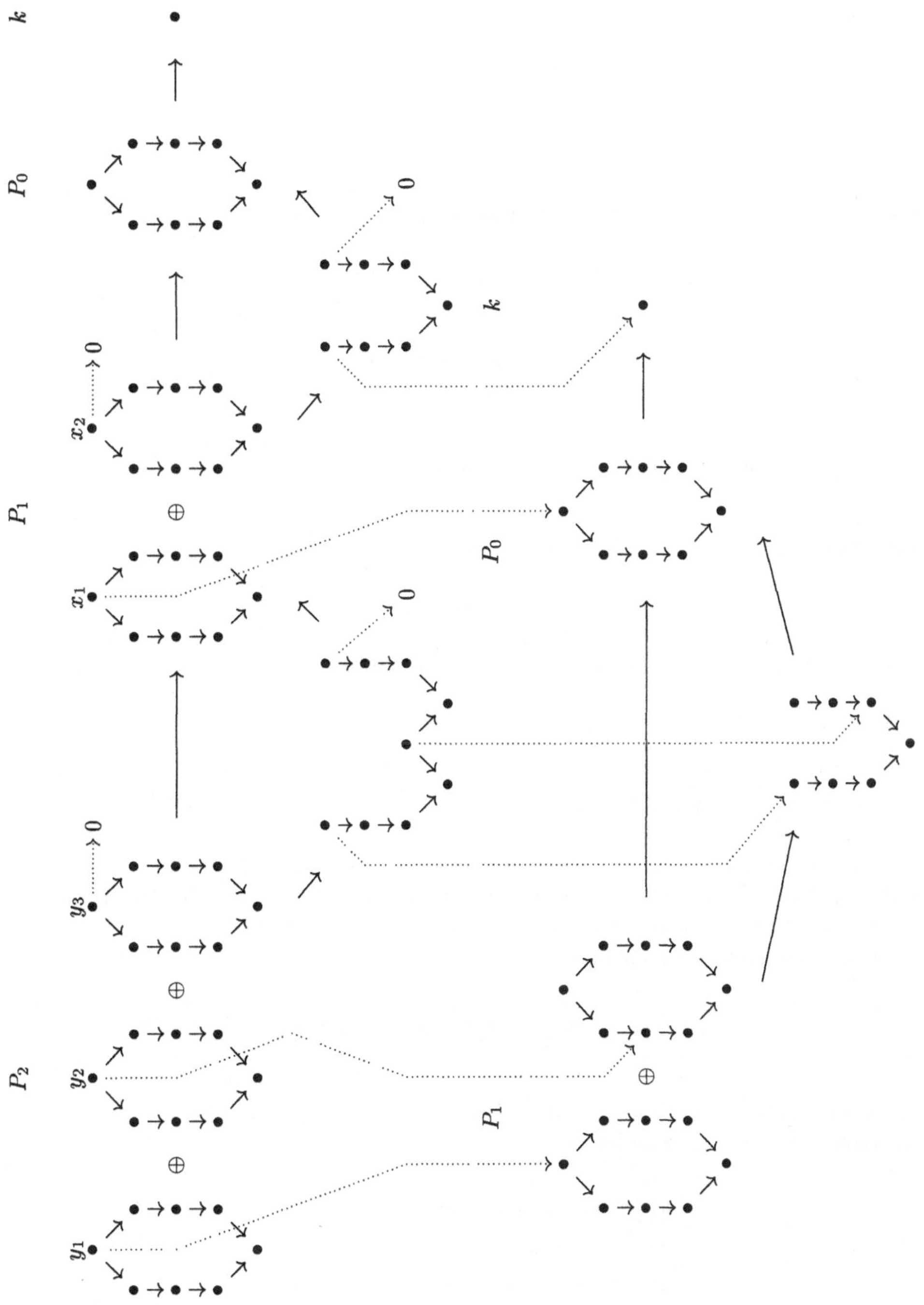

k
P_0
P_1
P_2
x_1
x_2
y_1
y_2
y_3
0
k
P_0
P_1

Analogously, we have b_1 represented by

$$\begin{array}{rcl} b_1 \ : \ P_1 & \longrightarrow & k \\ x_1 & \longmapsto & 0 \\ x_2 & \longmapsto & 1 \end{array}$$

The diagram reveals that $a_1 b_1 = 0$. Finally, c_2 is given by

$$\begin{array}{rcl} c_2 \ : \ P_2 & \longrightarrow & k \\ y_1 & \longmapsto & 0 \\ y_2 & \longmapsto & 1 \\ y_3 & \longrightarrow & 0 \end{array}$$

Continuing along these lines we can prove the following theorem.

Theorem 7.8 *If* $\operatorname{char} k = 2$, *then*

$$H^*(D_8, k) = k[a_1, b_1, c_2]/(a_1 b_1)$$

where a_1 *and* b_1 *are in degree* 1, *and* c_2 *is in degree* 2.

Our next example is

$$G = A_4 = \langle x, y, z \mid x^2 = y^2 = (xy)^2 = z^3 = 1, zxz^{-1} = y, zyz^{-1} = xy \rangle,$$

the alternating group on 4 letters. Here, for example, $x = (1\ 2)(3\ 4)$, $y = (1\ 4)(2\ 3)$, and $z = (1\ 2\ 3)$. We assume that k has characteristic 2 and that it contains a quadratic extension of its prime field. So we have $\omega \in k$ with $1 + \omega + \omega^2 = 0$, that is, ω is a primitive 3rd root of 1. Let

$$e_i := \sum_{j=0}^{2} \omega^{-ij} z^j \quad \text{for } i = 0, 1, 2.$$

These elements are orthogonal idempotents ($e_i e_j = \delta_{i,j} e_i$) and $e_0 + e_1 + e_2 = 1$. Moreover $z e_i = \omega^i e_i$. Now let

$$\begin{aligned} u_1 &:= x + \omega^2 y + \omega xy, \\ u_2 &:= x + \omega y + \omega^2 xy, \\ u_3 &:= 1 + x + y + xy = u_1 u_2 = u_2 u_1. \end{aligned}$$

Note that $z u_i z^{-1} = \omega^i u_i$.

The indecomposable projectives look like

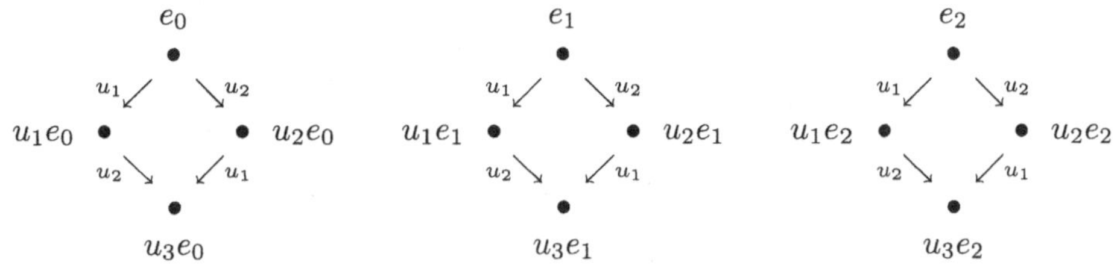

Here we have three simple modules, which we denote by their z-eigenvalues, namely, 1, ω, and ω^2.

At this point it is appropriate to introduce a new style of diagram. We let the vertices of the diagram be simple kG-modules, and the edges (or arrows) are extension classes between the vertices. That is, an edge is a class in $\mathrm{Ext}^1_{kG}(M,N)$ where M and N are the vertices. So the indecomposable projectives for kA_4 look like

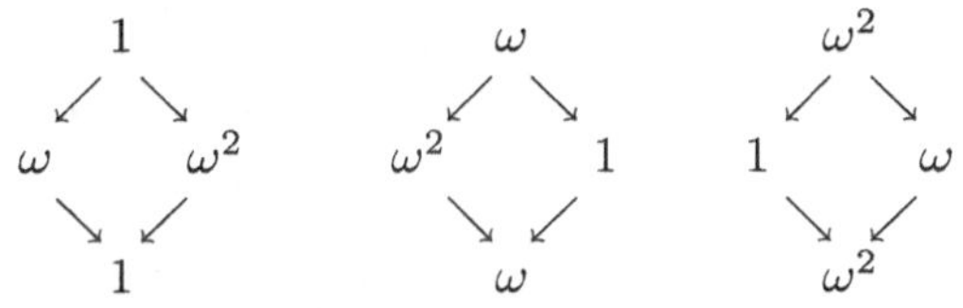

kA_4e_0 $\qquad$ kA_4e_1 $\qquad$ kA_4e_2

Then it can be checked that a minimal projective resolution for $G = A_4$ is

$$\begin{aligned}\cdots \to kGe_0 \oplus kGe_1 \oplus kGe_2 \oplus kGe_0 &\to kGe_2 \oplus kGe_0 \oplus kGe_1 \\ &\to kGe_1 \oplus kGe_2 \to kGe_0 \to k \to 0\end{aligned}$$

and one can deduce that the cohomology ring is given as described by the next theorem.

Theorem 7.9 *If* $\mathbb{F}_4 \hookrightarrow k$, *then*

$$H^*(A_4, k) = k[a_2, b_3, c_3]/(a_2^3 + b_3c_3)$$

where the indices denote the degrees of the elements.

Next take $G = S_4$, $p = 2$. Then kG has two simple modules, k and M with $\dim M = 2$. The indecomposable projectives look like

$$
\begin{array}{ccc}
 & k & \\
\swarrow & & \searrow \\
k & & M \\
\downarrow & & \downarrow \\
M & & k \\
\searrow & & \swarrow \\
 & k &
\end{array}
\qquad\qquad
\begin{array}{ccc}
 & M & \\
\swarrow & & \searrow \\
k & & \\
\downarrow & & M \\
k & & \\
\searrow & & \swarrow \\
 & M &
\end{array}
$$

Theorem 7.10 *If* $\operatorname{char} k = 2$, *then*

$$H^*(S_4, k) = k[a_1, b_2, c_3]/(a_1 c_3)$$

where the indices denote the degrees of the elements.

As a last example we take $G = A_5$, $p = 2$, and $\mathbb{F}_4 \hookrightarrow k$. Then the principal block of kG has three simple modules, k, M_1, and M_2 of dimensions 1, 2, and 2, respectively. The indecomposable projective modules look like

$$
\begin{array}{ccc}
 & k & \\
\swarrow & & \searrow \\
M_1 & & M_2 \\
\downarrow & & \downarrow \\
k & & k \\
\downarrow & & \downarrow \\
M_2 & & M_1 \\
\searrow & & \swarrow \\
 & k &
\end{array}
\qquad\qquad
\begin{array}{c}
M_1 \\ \downarrow \\ k \\ \downarrow \\ M_2 \\ \downarrow \\ k \\ \downarrow \\ M_1
\end{array}
\qquad\qquad
\begin{array}{c}
M_2 \\ \downarrow \\ k \\ \downarrow \\ M_1 \\ \downarrow \\ k \\ \downarrow \\ M_2
\end{array}
$$

Theorem 7.11 *If* $\mathbb{F}_4 \hookrightarrow k$, *then*

$$H^*(A_5, k) = k[a_2, b_3, c_3]/(a_2^3 + b_3 c_3)$$

where the indices denote the degrees of the elements.

For further information on diagrams and cohomology see the book by Benson [B1] and the paper [BC].

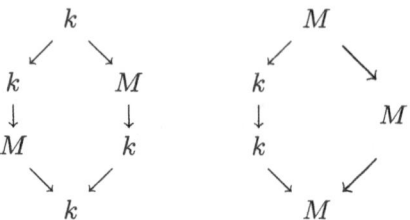

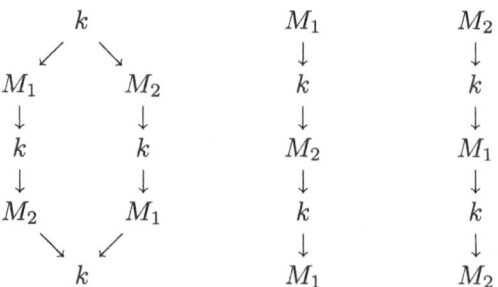

8 Relative projectivity

Throughout this section V denotes a module in ${}_{kG}\mathfrak{mod}$.

In this section we develop techniques and results concerned with projectivity relative to a module. We shall see that this is a generalization of the more standard projectivity relative to a subgroup or set of subgroups. The latter plays a role in the theory of blocks for group algebras and so has a vast literature. The projectivity relative to a module was first seen in an unpublished manuscript of Okuyama [O]. Its definition, however, is just a special case of the relative homological algebra that can be defined for a projective class of epimorphisms [HS, Chap. 10, Sect. 1] or for a pair of adjoint exact functors [HS, Chap. 10, Sect. 4].

The reason for our interest in the subject will become clear in the next section, where we will consider the special case of projectivity relative to a tensor product of L_ζ's for $\zeta \in H^n(G,k) - \{0\}$, $n \geqslant 1$. We will show how to represent ideals in the cohomology ring by exact functors which come from relative projective resolutions. We begin with some definitions and easy results.

Definition A module M in ${}_{kG}\mathfrak{mod}$ is termed *V-projective* or *relatively V-projective* if $\Omega^0(M)$ is isomorphic to a direct summand of $V \otimes N$ for some suitable module N in ${}_{kG}\mathfrak{mod}$. The module M is called *V-injective* or *relatively V-injective* if and only if it is V-projective.

Let $\mathcal{P}(V)$ denote the collection of all V-projective modules in ${}_{kG}\mathfrak{mod}$.

An exact sequence $E:\ 0 \longrightarrow A \xrightarrow{\alpha} B \xrightarrow{\beta} C \longrightarrow 0$ in ${}_{kG}\mathfrak{mod}$ is said to be *V-split* if $V \otimes E:\ 0 \longrightarrow V \otimes A \xrightarrow{1\otimes\alpha} V \otimes B \xrightarrow{1\otimes\beta} V \otimes C \longrightarrow 0$ splits.

Remark If M is not projective, then M is V-projective if and only if $M \mid V \otimes N$ for some module N in ${}_{kG}\mathfrak{mod}$.

Proposition 8.1

(i) *If* $U \in \mathcal{P}(V)$ *then* $\mathcal{P}(U) \subseteq \mathcal{P}(V)$.

(ii) $\mathcal{P}(V) = \mathcal{P}(V^*)$.

(iii) *For any integer* n, $\mathcal{P}(V) = \mathcal{P}(\Omega^n(V))$.

(iv) *If* $p \nmid \dim V$, *then* $\mathcal{P}(V) = {}_{kG}\mathfrak{mod}$.

(v) *Let* $\mathcal{P}$ $\big(= \mathcal{P}(0)\big)$ *be the collection of all projective modules. Then* $\mathcal{P} \subseteq \mathcal{P}(V)$, *and if* V *is projective, then* $\mathcal{P} = \mathcal{P}(V)$.

Proof. (i) is obvious; in fact, $\Omega^0(U) \mid V \otimes A$ $\big($hence $U \mid (V \otimes A) \oplus (\text{proj})\big)$ implies $\Omega^0(M) \mid U \otimes B \Longrightarrow \Omega^0(M) \mid V \otimes A \otimes B$ because $\Omega^0(M)$ is projective-free.

(ii) Recall that $V \mid V \otimes V^* \otimes V$ (Proposition 2.3). So V is V^*-projective and dually V^* is V-projective, whence $\mathcal{P}(V) = \mathcal{P}(V^*)$ by (i).

(iii) Proposition 4.4 shows that $\Omega^n(V) \mid V \otimes \Omega^n(k)$ and $\Omega^0(V) \mid \Omega^n(V) \otimes \Omega^{-n}(k)$. So (iii) holds.

(iv) follows from Lemma 2.2.

(v) Both statements are obvious: $0 \in \mathcal{P}(V)$, and if V is projective, then $\Omega^0(V) = 0$. □

Note that (iv) gives a weak version of Corollary 1.6.

Corollary 8.2 *If $p \mid |G|$ and V is projective, then $p \mid \dim V$.*

Proposition 8.3 *If $E:\ 0 \longrightarrow A \longrightarrow B \longrightarrow C \longrightarrow 0$ is V-split and M is V-projective, then $M \otimes E:\ 0 \longrightarrow M \otimes A \longrightarrow M \otimes B \longrightarrow M \otimes C \longrightarrow 0$ splits.*

Proof. $\Omega^0(M) \mid V \otimes N$ for some module N in ${}_{kG}\mathfrak{mod}$. Hence the exact sequence $\Omega^0(M) \otimes E$ is a direct summand of $V \otimes N \otimes E$, which splits. So $M \otimes E$ splits. □

The above proof is a bit more complicated than what we have written down. The problem is that it depends on the naturality of both the tensor product and the splitting. Another way of looking at it is the following. Suppose that $V^* \otimes N = M \oplus M'$. Now we know that the class $\mathrm{class}(V^* \otimes N \otimes E) = \mathrm{id}_{V^* \otimes N} \cdot \mathrm{class}(E) \in \mathrm{Ext}^1_{kG}(V^* \otimes N \otimes C, V^* \otimes N \otimes A)$ is zero. Then so also is its projection to $\mathrm{Ext}^1_{kG}(M \otimes C, M \otimes A)$ which is $\mathrm{class}(M \otimes E)$.

Corollary 8.4 *Let $E:\ 0 \longrightarrow A \longrightarrow B \longrightarrow C \longrightarrow 0$ be a short exact sequence in ${}_{kG}\mathfrak{mod}$. The following statements are equivalent.*

(i) *E is V-split.*

(ii) *E is V^*-split.*

(iii) *E is $\Omega^n(V)$-split (for any integer n).*

Proposition 8.5 *Suppose that we have a diagram*

$$\begin{array}{ccccccccc} & & & & & & M & & \\ & & & & & & \downarrow{\scriptstyle\theta} & & \\ 0 & \longrightarrow & A & \longrightarrow & B & \xrightarrow{\beta} & C & \longrightarrow & 0 \end{array}$$

with exact V-split row and with M V-projective. Then there exists a kG-homomorphism $\mu : M \longrightarrow B$ with $\beta\mu = \theta$.

Proof. We have the following commutative diagram.

$$\begin{array}{ccc} \mathrm{Hom}_{kG}(M,B) & \xrightarrow{\beta_*} & \mathrm{Hom}_{kG}(M,C) \\ \cong & & \cong \\ \mathrm{Hom}_{kG}(k, M^* \otimes B) & \xrightarrow{(1\otimes\beta)_*} & \mathrm{Hom}_{kG}(k, M^* \otimes C) \end{array}$$

Since M is V-projective, M^* is V^*-projective, hence V-projective by Proposition 8.1 (ii). So $0 \to M^*\otimes A \to M^*\otimes B \xrightarrow{1\otimes\beta} M^*\otimes C \to 0$ splits. Therefore $(1\otimes\beta)_*$ is surjective and so is β_*. So there exists $\mu \in \mathrm{Hom}_{kG}(M,B)$ with $\theta = \beta_*(\mu) = \beta\mu$. □

Remark The proposition above has a dual counterpart. That is, if we are given a diagram

$$\begin{array}{ccccccccc} 0 & \longrightarrow & A & \xrightarrow{\alpha} & B & \longrightarrow & C & \longrightarrow & 0 \\ & & \downarrow{\scriptstyle\psi} & & & & & & \\ & & M & & & & & & \end{array}$$

with exact V-split row and with M V-injective, i. e., V-projective, then there exists a kG-homomorphism $\varphi : B \to M$ with $\varphi\alpha = \psi$.

Definition A *V-projective resolution* of a module M in ${}_{kG}\mathfrak{mod}$ is a nonnegative complex P_* of V-projective modules together with a kG-homomorphism $P_0 \overset{\varepsilon}{\twoheadrightarrow} M$ such that the sequence

$$P_* \overset{\varepsilon}{\twoheadrightarrow} M : \qquad \cdots \to P_2 \xrightarrow{\partial_2} P_1 \xrightarrow{\partial_1} P_0 \xrightarrow{\varepsilon} M \to 0$$

is exact and totally V-split.

Totally V-split means that all of the exact sequences

$$0 \longrightarrow \ker\varepsilon \longrightarrow P_0 \xrightarrow{\varepsilon} M \longrightarrow 0,$$

$$0 \longrightarrow \ker\partial_i \longrightarrow P_i \xrightarrow{\partial_i} \mathrm{im}\,\partial_i \longrightarrow 0 \qquad (i = 1, 2, \dots)$$

are V-split.

Similarly, there is the notion of a *V-injective resolution* $M \overset{\theta}{\hookrightarrow} Q_*$.

Lemma 8.6 *The exact sequence*

$$0 \longrightarrow \ker\mathrm{Tr} \longrightarrow V\otimes V^* \xrightarrow{\mathrm{Tr}} k \longrightarrow 0$$

is V-split.

Proof. This follows from Proposition 2.3 together with Lemma 6.12. □

Proposition 8.7 *Every module M in ${}_{kG}\mathfrak{mod}$ has a V-projective resolution.*

Proof. The sequences

$$0 \to \ker \mathrm{Tr} \otimes M \to V \otimes V^* \otimes M \to M \to 0,$$
$$0 \to \ker \mathrm{Tr} \otimes \ker \mathrm{Tr} \otimes M \to V \otimes V^* \otimes \ker \mathrm{Tr} \otimes M \to \ker \mathrm{Tr} \otimes M \to 0,$$
$$\vdots$$

are V-split. So by splicing these sequences together we get a V-projective resolution of M:

$$\begin{array}{ccccccc} \cdots & \longrightarrow & V \otimes V^* \otimes \ker \mathrm{Tr} \otimes M & \longrightarrow & V \otimes V^* \otimes M \to M \to 0 \\ & \searrow \nearrow & & \searrow \nearrow & \\ (\ker \mathrm{Tr})^{\otimes 2} \otimes M & & & \ker \mathrm{Tr} \otimes M & \end{array}$$

If M is V-projective we can do better, namely,

$$0 \longrightarrow M \xrightarrow{\mathrm{id}_M} M \longrightarrow 0$$

is a V-projective resolution. □

Theorem 8.8 (Comparison Theorem) *Let $P_* \overset{\varepsilon}{\twoheadrightarrow} M$ and $Q_* \overset{\theta}{\twoheadrightarrow} M$ be two V-projective resolutions of M in ${}_{kG}\mathfrak{mod}$. Then there is a chain map*

$$\mu_* : (P_* \overset{\varepsilon}{\twoheadrightarrow} M) \to (Q_* \overset{\theta}{\twoheadrightarrow} M)$$

which lifts the identity on M.

Proof. Assume by induction that we have constructed $\mu_0, \dots, \mu_{i-1}$, so that we have a commutative diagram

$$\begin{array}{ccccccccccccc} \cdots & \longrightarrow & P_i & \longrightarrow & P_{i-1} & \xrightarrow{\partial^P_{i-1}} & \cdots & \xrightarrow{\partial^P_1} & P_0 & \xrightarrow{\varepsilon} & M & \longrightarrow & 0 \\ & & & & \downarrow{\scriptstyle \mu_{i-1}} & & & & \downarrow{\scriptstyle \mu_0} & & \| & & \\ \cdots & \longrightarrow & Q_i & \longrightarrow & Q_{i-1} & \xrightarrow{\partial^Q_{i-1}} & \cdots & \xrightarrow{\partial^Q_1} & Q_0 & \xrightarrow{\theta} & M & \longrightarrow & 0 \end{array}$$

Then μ_{i-1} restricted to $\ker \partial^P_{i-1} \cong \operatorname{im} \partial^P_i$ induces $\bar{\mu}_{i-1} : \operatorname{im} \partial^P_i \to \operatorname{im} \partial^Q_i$. So we have a commutative diagram

$$\begin{array}{ccccccccc} 0 & \longrightarrow & \ker \partial^P_i & \longrightarrow & P_i & \longrightarrow & \operatorname{im} \partial^P_i & \longrightarrow & 0 \\ & & & & \downarrow{\scriptstyle \mu_i} & & \downarrow{\scriptstyle \bar{\mu}_{i-1}} & & \\ 0 & \longrightarrow & \ker \partial^Q_i & \longrightarrow & Q_i & \longrightarrow & \operatorname{im} \partial^Q_i & \longrightarrow & 0 \end{array}$$

We get μ_i because P_i is V-projective and the bottom row is V-split. □

Proposition 8.9 *Every module M in ${}_{kG}\mathfrak{mod}$ has a minimal V-projective resolution and a minimal V-injective resolution.*

The last proposition can be proved using arguments which are very similar to those in the proof of the existence of minimal projective resolutions.

Definition For M in ${}_{kG}\mathfrak{mod}$ and $n \in \mathbb{Z}$ define $\Omega^n_V(M)$ as in the definition of $\Omega^n(M)$ on page 14, but using V-projective and V-injective resolutions instead of projective and injective resolutions. $\Omega^0_V(M) = \Omega^n_V(\Omega^{-n}_V(M))$ is the sum of the non-V-projective direct summands of V.

Exercise 8.1 Show that a V-projective resolution $P_* \xrightarrow{\varepsilon} M$ is minimal if and only if for all $n > 0$, $\operatorname{im}\partial_n$ has no nonzero V-projective direct summands.

A consequence of Exercise 8.1 is the following lemma.

Lemma 8.10 $\Omega^n_V(M \oplus N) \cong \Omega^n_V(M) \oplus \Omega^n_V(N)$. *More generally, the direct sum of two minimal V-projective resolutions is a minimal V-projective resolution for the direct sum of the modules.*

Lemma 8.11 *If M is indecomposable, then so is $\Omega^n_V(M)$ for any n for which $\Omega^n_V(M) \neq 0$.*

Proof. $\Omega^n_V(M) \cong A \oplus B$ implies that $M \cong \Omega^{-n}_V(A \oplus B) \oplus (V\text{-proj}) \cong \Omega^{-n}_V(A) \oplus \Omega^{-n}_V(B) \oplus (V\text{-proj})$. □

A few notes on tensor products will be of value later. The first is very easy to see.

Proposition 8.12 *If $P_* \xrightarrow{\varepsilon} M$ is a V-projective resolution of M in ${}_{kG}\mathfrak{mod}$, then for any module N in ${}_{kG}\mathfrak{mod}$, $P_* \otimes N \xrightarrow{\varepsilon\otimes 1} M \otimes N$ is a V-projective resolution.*

Corollary 8.13 *Let V_k be a V-projective cover of the trivial module k. Then $\mathcal{P}(V) = \mathcal{P}(V_k)$.*

Proof. Clearly V_k is V-projective. But by definition $0 \longrightarrow \ker\varepsilon \longrightarrow V_k \xrightarrow{\varepsilon} k \longrightarrow 0$ is V-split. So V is a direct summand of $V_k \otimes V$ and hence V is V_k-projective. □

Proposition 8.14 *If $P^{(1)}_* \xrightarrow{\varepsilon_1} M_1$ and $P^{(2)}_* \xrightarrow{\varepsilon_2} M_2$ are V_1- and V_2-projective resolutions of M_1 and M_2, respectively, then $P^{(1)}_* \otimes P^{(2)}_* \xrightarrow{\varepsilon_1\otimes\varepsilon_2} M_1 \otimes M_2$ is a $V_1 \otimes V_2$-projective resolution.*

Proof. It is easy to see that $P_i^{(1)} \otimes P_j^{(2)}$ is $V_1 \otimes V_2$-projective for any i and j. So we need only notice that $V_1 \otimes V_2 \otimes P_*^{(1)} \otimes P_*^{(2)} \cong (V_1 \otimes P_*^{(1)}) \otimes (V_2 \otimes P_*^{(2)})$ is totally split. □

9 Relative projectivity and ideals in cohomology

In this section we investigate methods of representing ideals in cohomology by exact sequences. Parts of the section will appear in joint work with Wayne Wheeler [CW] and with Chuang Peng [CP]. The foundation for our investigation was laid in the results at the end of Sect. 6. The point is that we will characterize ideals in cohomology by the modules in ${}_{kG}\mathfrak{mod}$ whose cohomology the ideals annihilate. To make the investigation more meaningful we begin by quoting the theorem on finite generation of Evens and Venkov (see [E]). One of the main facts which we use is that the ideals in cohomology are finitely generated.

Theorem 9.1 (Evens, Venkov) *$H^*(G,k)$ is a finitely generated k-algebra. Moreover if M and N are in ${}_{kG}\mathfrak{mod}$, then $\operatorname{Ext}^*_{kG}(M,N)$ is a finitely generated module over $H^*(G,k)$.*

Remark Of course, as everywhere in this text, G is a finite group. But the theorem holds if the coefficient ring k is any commutative Noetherian ring. In particular, Theorem 9.1 says that $H^*(G,k)$ is a Noetherian ring, and hence its ideals are finitely generated.

If $p = 2$, then $H^*(G,k) \cong k[x_1, \ldots, x_n]/I$ where we may choose the finitely many generators $x_1, \ldots, x_n$ to be homogeneous elements (of various degrees). The ideal I is then homogeneous, i. e., it is generated by the homogeneous elements that it contains. For $p > 2$ the elements of odd degree anticommute. So we have $H^*(G,k) \cong k[x_1, \ldots, x_n] \otimes \Lambda(y_1, \ldots, y_m)/I$ in this case, where $x_1, \ldots, x_n$ have even degrees and $y_1, \ldots, y_m$ have odd degrees. The ideal I is again homogeneous.

Definition For a module M in ${}_{kG}\mathfrak{mod}$, let $J(M)$ be the ideal in $H^*(G,k)$ generated by all homogeneous elements of positive degree which annihilate the cohomology of M. That is, if $\zeta \in H^n(G,k)$, $n > 0$, then $\zeta \in J(M)$ if and only if ζ annihilates $\widehat{\operatorname{Ext}}^*_{kG}(M,N)$ and $\widehat{\operatorname{Ext}}^*_{kG}(N,M)$ for any module N in ${}_{kG}\mathfrak{mod}$.

We have noted before that a homogeneous element ζ of positive degree is in $J(M)$ if and only if $\zeta \cdot \mathrm{id}_M = 0$ where id_M is the identity element in $\widehat{\operatorname{Ext}}^0_{kG}(M,M)$. Thus for $M \neq 0$, $J(M)$ is the annihilator in $H^*(G,k)$ of $\operatorname{Ext}^*_{kG}(M,M)$, and an element $\zeta \in H^n(G,k)$, $n > 0$, is in $J(M)$ if and only if ζ annihilates the ordinary cohomology $\operatorname{Ext}^*_{kG}(M,N)$ and $\operatorname{Ext}^*_{kG}(N,M)$ for any kG-module N.

Our first result (the next lemma) is obvious in light of what we have done so far. However, a lot of what we do in this section is a generalization of the result.

Lemma 9.2 *Let $\zeta \in H^n(G,k) - \{0\}$, $n \geqslant 1$, and let—cf.* (6.5) *on page 39—*

$$E_\zeta : \quad 0 \to k \to \Omega^{-1}(L_\zeta) \to \Omega^{n-1}(k) \to 0$$

be a sequence which represents $\zeta \in H^n(G,k) = \operatorname{Ext}^n_{kG}(k,k) \cong \operatorname{Ext}^1_{kG}(\Omega^{n-1}(k),k)$. Then E_ζ represents the ideal $(\zeta) \subseteq H^*(G,k)$ in the sense that $(\zeta) \subseteq J(M)$ if and only if $E_\zeta \otimes M$ splits. Moreover $E_\zeta \otimes M$ splits if and only if $L_\zeta \otimes M \cong \Omega^n(M) \oplus \Omega(M) \oplus (\mathrm{proj})$.

Proof. $E_\zeta \otimes M$ represents $\zeta \cdot \mathrm{id}_M$. The last statement follows from Lemma 6.12. □

Definition An element $\zeta \in H^n(G,k) - \{0\}$, $n \geqslant 1$, is *productive* provided $\zeta \in J(L_\zeta)$.

We know from Theorem 6.11 that if $p > 2$, then all $\zeta \in H^n(G,k) - \{0\}$, $n \geqslant 1$, n even, are productive.

Proposition 9.3 *Let $\zeta \in H^n(G,k)$ be a productive element. Then a minimal L_ζ-projective resolution of k has the form*

$$\cdots \to \begin{matrix}\Omega^{2-3n}(L_\zeta)\\ \oplus \\ (\mathrm{proj})\end{matrix} \xrightarrow{\partial_2} \begin{matrix}\Omega^{1-2n}(L_\zeta)\\ \oplus \\ (\mathrm{proj})\end{matrix} \xrightarrow{\partial_1} \begin{matrix}\Omega^{-n}(L_\zeta)\\ \oplus \\ (\mathrm{proj})\end{matrix} \to k \to 0.$$

Moreover $\Omega^i_{L_\zeta}(k) = \operatorname{im} \partial_i \cong \Omega^{i(1-n)}(k)$.

Proof. Begin with $E_\zeta : 0 \to k \to \Omega^{-1}(L_\zeta) \to \Omega^{n-1}(k) \to 0$. Since ζ is productive, E_ζ is L_ζ-split by Lemma 9.2. Translating by Ω^{1-n} we get a sequence

$$0 \to \Omega^{1-n}(k) \to \Omega^{-n}(L_\zeta) \oplus (\mathrm{proj}) \to k \to 0$$

which is also L_ζ-split. Moreover $\Omega^{-n}(L_\zeta) \oplus (\mathrm{proj})$ is L_ζ-projective. By further translating by Ω^{1-n} we get L_ζ-split exact sequences

$$0 \longrightarrow \Omega^{2-2n}(k) \longrightarrow \Omega^{1-2n}(L_\zeta) \oplus (\mathrm{proj}) \longrightarrow \Omega^{1-n}(k) \longrightarrow 0,$$

$$0 \longrightarrow \Omega^{3-3n}(k) \longrightarrow \Omega^{2-3n}(L_\zeta) \oplus (\mathrm{proj}) \longrightarrow \Omega^{2-2n}(k) \longrightarrow 0,$$

etc., all with L_ζ-projective middle terms. So splicing these together we get the L_ζ-projective resolution

$$\cdots \to \begin{matrix}\Omega^{2-3n}(L_\zeta)\\ \oplus \\ (\mathrm{proj})\end{matrix} \xrightarrow{\partial_2} \begin{matrix}\Omega^{1-2n}(L_\zeta)\\ \oplus \\ (\mathrm{proj})\end{matrix} \xrightarrow{\partial_1} \begin{matrix}\Omega^{-n}(L_\zeta)\\ \oplus \\ (\mathrm{proj})\end{matrix} \to k \to 0$$

as desired. The minimality of the resolution follows from the fact that $\Omega^{i(1-n)}(k)$ is indecomposable. □

Now suppose that we are given homogeneous elements $\zeta_i \in H^{n_i}(G,k) - \{0\}$, $n_i \geqslant 1$, for $i = 1, \ldots, n$. For each i we have an exact sequence

$$0 \longrightarrow \Omega^{n_i}(k) \xrightarrow{\hat{\zeta}_i} k \oplus (\text{proj}) \xrightarrow{\sigma_i} \Omega^{-1}(L_{\zeta_i}) \longrightarrow 0$$

where $\hat{\zeta}_i$ represents $\zeta_i \in \underline{\text{Hom}}_{kG}(\Omega^{n_i}(k), k)$.

Since each σ_i is surjective, so is the tensor product $\sigma_1 \otimes \cdots \otimes \sigma_n$. We get an exact sequence

$$0 \to U(\zeta_1, \ldots, \zeta_n) \xrightarrow{\theta(\zeta_1,\ldots,\zeta_n)} k \oplus (\text{proj}) \xrightarrow{\sigma_1 \otimes \cdots \otimes \sigma_n} \bigotimes_{i=1}^{n} \Omega^{-1}(L_{\zeta_i}) \to 0 \quad (9.1)$$

which defines the module $U(\zeta_1, \ldots, \zeta_n)$. Note that the σ_i's are well-defined up to the addition of homomorphisms which factor through projectives. Hence the isomorphism class of $U(\zeta_1, \ldots, \zeta_n)$, being the third object in the triangle of the morphism $\sigma_1 \otimes \cdots \otimes \sigma_n$, is well-defined in the stable category.

There is another way of defining the module $U(\zeta_1, \ldots, \zeta_n)$. Let $\mathcal{C}(\zeta_i)_*$ be the complex

$$\mathcal{C}(\zeta_i)_* : \qquad 0 \to \Omega^{n_i}(k) \xrightarrow{\hat{\zeta}_i} k \oplus (\text{proj}) \to 0 \qquad \text{with } k \oplus (\text{proj}) \text{ in degree } 0.$$

Then $H_*(\mathcal{C}(\zeta_i)_*) = H_0(\mathcal{C}(\zeta_i)_*) \cong \Omega^{-1}(L_{\zeta_i})$. Take the tensor product

$$\mathcal{C}_* := \bigotimes_{i=1}^{n} \mathcal{C}(\zeta_i)_* : \qquad 0 \to \mathcal{C}_n \to \cdots \to \mathcal{C}_0 \to 0$$

with

$$\begin{aligned} \mathcal{C}_0 &\cong k \oplus (\text{proj}), \\ \mathcal{C}_1 &\cong \bigoplus_{i=1}^{n} \Omega^{n_i}(k) \oplus (\text{proj}), \\ &\vdots \\ \mathcal{C}_n &\cong \Omega^{\sum_{i=1}^{n} n_i}(k) \oplus (\text{proj}), \end{aligned}$$

and $H_*(\mathcal{C}_*) = H_0(\mathcal{C}_*) \cong \bigotimes_{i=1}^{n} \Omega^{-1}(L_{\zeta_i})$. We define $\mathcal{D}_*$ to be the truncated complex

$$\mathcal{D}_* : \qquad 0 \to \mathcal{D}_{n-1} \to \cdots \to \mathcal{D}_0 \to 0$$

where $\mathcal{D}_i = \mathcal{C}_{i+1}$ and for $i \neq 0$ the differential $\mathcal{D}_i \xrightarrow{\partial} \mathcal{D}_{i-1}$ is the same as $\mathcal{C}_{i+1} \xrightarrow{\partial} \mathcal{C}_i$.

Lemma 9.4 $H_*(\mathcal{D}_*) = H_0(\mathcal{D}_*) \cong U(\zeta_1, \ldots, \zeta_n)$.

Proof. The first row in the following diagram is exact.

$$\begin{array}{ccccccc} 0 \longrightarrow & \mathcal{C}_n \to \cdots \to \mathcal{C}_1 & \longrightarrow & \mathcal{C}_0 & \xrightarrow{\sigma_1\otimes\cdots\otimes\sigma_n} & \bigotimes_{i=1}^{n} \Omega^{-1}(L_{\zeta_i}) \to 0 \\ & \| \qquad\quad \| & & & & \\ 0 \longrightarrow & \mathcal{D}_{n-1} \to \cdots \to \mathcal{D}_0 & \longrightarrow & 0 & & \end{array}$$

It follows that $H_0(\mathcal{D}_*) \cong \operatorname{im}(\mathcal{C}_1 \xrightarrow{\partial} \mathcal{C}_0) \cong \ker(\sigma_1 \otimes \cdots \otimes \sigma_n) = U(\zeta_1, \ldots, \zeta_n)$. □

Now we have a triangle shift of the sequence (9.1) as follows:

$$E(\zeta_1, \ldots, \zeta_n): \qquad 0 \to k \xrightarrow{\sigma} \bigotimes_{i=1}^{n} \Omega^{-1}(L_{\zeta_i}) \to \Omega^{-1}\big(U(\zeta_1, \ldots, \zeta_n)\big) \to 0.$$

Theorem 9.5 *The sequence*

$$E(\zeta_1, \ldots, \zeta_n): \quad 0 \to k \xrightarrow{\sigma} \bigotimes_{i=1}^{n} \Omega^{-1}(L_{\zeta_i}) \to \Omega^{-1}\big(U(\zeta_1, \ldots, \zeta_n)\big) \to 0$$

represents the ideal $I := (\zeta_1, \ldots, \zeta_n)$ *in the sense that, for any module* M *in* ${}_{kG}\mathfrak{mod}$, $E(\zeta_1, \ldots, \zeta_n) \otimes M$ *splits if and only if* $I \subseteq J(M)$.

Proof. For each i we have a commutative diagram

$$\begin{array}{cccccc} E_i: & 0 \to k & \xrightarrow{\sigma_i'} & \Omega^{-1}(L_{\zeta_i}) & \to & \Omega^{n_i-1}(k) \to 0 \\ & \| & & \downarrow{\scriptstyle \mu_i} & & \downarrow{\scriptstyle \nu_i} \\ E(\zeta_1, \ldots, \zeta_n): & 0 \to k & \xrightarrow{\sigma_1'\otimes\cdots\otimes\sigma_n'} & \bigotimes_{i=1}^{n} \Omega^{-1}(L_{\zeta_i}) & \to & \Omega^{-1}\big(U(\zeta_1, \ldots, \zeta_n)\big) \to 0 \end{array}$$

where $\mu_i = \sigma_1' \otimes \cdots \otimes \sigma_{i-1}' \otimes 1 \otimes \sigma_{i+1}' \otimes \cdots \otimes \sigma_n'$ and ν_i is induced from μ_i. So we see that E_i is the pullback of $E(\zeta_1, \ldots, \zeta_n)$ along ν_i. Therefore if $E(\zeta_1, \ldots, \zeta_n) \otimes M$ splits, then so does $E_i \otimes M$, and hence $\zeta_i \in J(M)$ in this case.

Now suppose that $\zeta_1, \ldots, \zeta_n \in J(M)$. Then, for each i, $E_i \otimes M$ splits. Let $\Omega^{-1}(L_{\zeta_i}) \otimes M \xrightarrow{\varphi_i} M$ be a splitting map for $M \xrightarrow{\sigma_i'\otimes 1} \Omega^{-1}(L_{\zeta_i}) \otimes M$. The composition

$$\Big(\bigotimes_{i=1}^{n} \Omega^{-1}(L_{\zeta_i})\Big) \otimes M \xrightarrow{1\otimes\varphi_n} \Big(\bigotimes_{i=1}^{n-1} \Omega^{-1}(L_{\zeta_i})\Big) \otimes M \xrightarrow{1\otimes\varphi_{n-1}} \cdots$$
$$\cdots \xrightarrow{1\otimes\varphi_2} \Omega^{-1}(L_{\zeta_1}) \otimes M \xrightarrow{\varphi_1} M$$

is a splitting map for $(1 \otimes \sigma_n' \otimes 1) \circ (1 \otimes \sigma_{n-1}' \otimes 1) \circ \cdots \circ (1 \otimes \sigma_2' \otimes 1) \circ (\sigma_1' \otimes 1) = \sigma_1' \otimes \cdots \otimes \sigma_n' \otimes 1$. □

Lemma 9.6 *If $\zeta \in J(M)$, then $\zeta \in J(M \otimes N)$ for any N.*

Proof. If $E_\zeta \otimes M$ splits, then so does $E_\zeta \otimes M \otimes N$. □

Proposition 9.7 *Suppose that $\zeta_1, \dots, \zeta_n$ are productive elements, and let $V := \bigotimes_{i=1}^{n} L_{\zeta_i}$. Then the sequence*

$$E(\zeta_1, \dots, \zeta_n): \qquad 0 \to k \xrightarrow{\sigma_1' \otimes \cdots \otimes \sigma_n'} \bigotimes_{i=1}^{n} \Omega^{-1}(L_{\zeta_i}) \to \Omega^{-1}(U(\zeta_1, \dots, \zeta_n)) \to 0$$

is the first step in a V-injective resolution of k.

Proof. Since ζ_i is productive, $\zeta_i \in J(L_{\zeta_i}) \subseteq J(V)$. So by Theorem 9.5, $E(\zeta_1, \dots, \zeta_n)$ is V-split. Also it is clear that $\bigotimes_{i=1}^{n} \Omega^{-1}(L_{\zeta_i})$ is V-projective. □

Proposition 9.8 *Suppose that $\zeta_1, \dots, \zeta_n$ are productive elements, and let $V := \bigotimes_{i=1}^{n} L_{\zeta_i}$. Then a module M in ${}_{kG}\mathfrak{mod}$ is V-projective if and only if $\zeta_1, \dots, \zeta_n \in J(M)$.*

Proof. If M is V-projective, then $E(\zeta_1, \dots, \zeta_n) \otimes M$ splits by the last proposition. So by Theorem 9.5, $\zeta_1, \dots, \zeta_n \in J(M)$.

The converse follows by the reverse argument. □

Definition An ideal $I \subseteq H^*(G, k)$ is called *productive* if it is generated by productive elements. Let $\zeta_1, \dots, \zeta_n$ be productive elements and $I := (\zeta_1, \dots, \zeta_n)$ the productive ideal they generate. A module M in ${}_{kG}\mathfrak{mod}$ is termed *I-projective* (or *I-injective*) if M is $\bigotimes_{i=1}^{n} L_{\zeta_i}$-projective. (The lesson of the last proposition is that I-projectivity is independent of the choice of productive generators.)

Now suppose that I is a productive ideal generated by the productive elements $\zeta_1, \dots, \zeta_n$. Let $E_I : 0 \to k \xrightarrow{\sigma_I} \Omega^{-1}(L_I) \to \Omega^{-1}(U_I) \to 0$ be the first step in a *minimal* I-injective resolution of k. So $U(\zeta_1, \dots, \zeta_n) \cong U_I \oplus U'$ where U' is I-projective. The module U_I is the object in the triangle

$$U_I \xrightarrow{\theta_I} k \xrightarrow{\sigma_I} \Omega^{-1}(L_I) \to \Omega^{-1}(U_I).$$

Lemma 9.9

(i) *The module U_I depends only on I (and not on the choice of productive generators) and θ_I is unique up to the addition of a map that factors through an I-projective module.*

(ii) *If $\zeta \in H^n(G,k)$ is a productive element, then we have $\big(U_{(\zeta)}, \theta_{(\zeta)}\big) = \big(\Omega^n(k), \zeta\big)$.*

Proof. Part (ii) and the independence statement in (i) have already been discussed. For the uniqueness we need only note that $\sigma_I : k \longrightarrow \Omega^{-1}(L_I)$ is an I-injective hull of k and then use the comparison theorem for I-injective modules. □

Proposition 9.10 *Let $I \subseteq J$ be productive ideals. Then there is a commutative diagram*

$$\begin{array}{ccc} U_I & \xrightarrow{\theta_I} & k \\ {\scriptstyle\theta_{I,J}}\downarrow & & \| \\ U_J & \xrightarrow{\theta_J} & k \end{array}$$

The map $\theta_{I,J}$ is unique up to a morphism which factors through an I-projective module.

Proof. We have $I \subseteq J \subseteq J(L_J)$ and hence L_J —and therefore also $\Omega^{-1}(L_J)$—are I-projective (or I-injective). So we get a commutative diagram

$$\begin{array}{ccccccccc} 0 & \longrightarrow & k & \longrightarrow & \Omega^{-1}(L_I) & \longrightarrow & \Omega^{-1}(U_I) & \longrightarrow & 0 \\ & & \| & & \downarrow{\scriptstyle\mu} & & \downarrow{\scriptstyle\nu} & & \\ 0 & \longrightarrow & k & \longrightarrow & \Omega^{-1}(L_J) & \longrightarrow & \Omega^{-1}(U_J) & \longrightarrow & 0 \end{array}$$

Here μ exists by the I-injectivity of $\Omega^{-1}(L_J)$. The map ν is induced by μ. Let $\theta_{I,J} := \Omega(\nu)$.

If we chose μ' instead of μ, then we would have the commutative diagram

$$\begin{array}{ccccccccc} 0 & \longrightarrow & k & \longrightarrow & \Omega^{-1}(L_I) & \longrightarrow & \Omega^{-1}(U_I) & \longrightarrow & 0 \\ & & \downarrow{\scriptstyle 0} & & \downarrow{\scriptstyle\mu-\mu'} & \swarrow & \downarrow{\scriptstyle\nu-\nu'} & & \\ 0 & \longrightarrow & k & \longrightarrow & \Omega^{-1}(L_J) & \longrightarrow & \Omega^{-1}(U_J) & \longrightarrow & 0 \end{array}$$

The uniqueness of $\theta_{I,J}$ then follows by general nonsense. □

10 Varieties and modules

We know that $H^*(G,k) = \operatorname{Ext}^*_{kG}(k,k)$ is a finitely generated k-algebra.

Definition Let $V_G(k)$ be the maximal ideal spectrum of $H^*(G,k)$.

So $V_G(k)$ is the set of all maximal ideals, topologized by the Zariski topology. That is, if I is an ideal in $H^*(G,k)$, then the set $V_G(I)$ of all maximal ideals which contain I is a closed set in the topology. Moreover every closed set has this form.

Example Suppose that the field k is algebraically closed. If G is an elementary abelian p-group, say $G \cong (\mathbb{Z}/p)^n$, then $H^*(G,k)/\operatorname{rad} H^*(G,k) \cong k[x_1,\dots,x_n]$ is a polynomial ring in n generators. So $V_G(k) \cong k^n$ in this case. For any $\alpha = (\alpha_1,\dots,\alpha_n) \in k^n$ there is a corresponding maximal ideal $\mathfrak{m}_\alpha$ in the polynomial ring, namely,

$$\mathfrak{m}_\alpha = \big\{ f(x_1,\dots,x_n) \in k[x_1,\dots,x_n] \bigm| f(\alpha) = 0 \big\}.$$

That is, $\mathfrak{m}_\alpha$ is the kernel of the homomorphism $k[x_1,\dots,x_n] \to k$ given by evaluation at the point α.

If $H \subseteq G$ is a subgroup, then we have a restriction homomorphism $H^*(G,k) \xrightarrow{\operatorname{res}_{G,H}} H^*(H,k)$. For suppose that $P_* \xrightarrow{\varepsilon} k$ is a projective kG-resolution of k. Then it is also a projective kH-resolution by restriction of coefficients. This is because kG is a projective kH-module. So we have an inclusion of complexes

$$\operatorname{Hom}_{kG}(P_*,k) \hookrightarrow \operatorname{Hom}_{kH}(P_*,k).$$

The induced map on cohomology is the restriction map.

Now if $\mathfrak{m} \in V_H(k)$ is a maximal ideal in $H^*(H,k)$, then its pullback

$$\operatorname{res}^{-1}_{G,H}(\mathfrak{m}) = \big\{ \zeta \in H^*(G,k) \bigm| \operatorname{res}_{G,H}(\zeta) \in \mathfrak{m} \big\}$$

is a maximal ideal in $H^*(G,k)$. Hence we have a map of varieties

$$\operatorname{res}^*_{G,H} : V_H(k) \to V_G(k)$$

sending $\mathfrak{m}$ to $\operatorname{res}^{-1}_{G,H}(\mathfrak{m})$.

The following theorem of Quillen has been fundamental for the development of the theory of varieties and cohomology rings. We do not give a proof here but rather refer the reader to Evens' book [E] for a readable treatment.

Theorem 10.1 (Quillen's Dimension Theorem) *Let $\mathcal{A}_p(G)$ be the set of all elementary abelian p-subgroups of G. Then*

$$V_G(k) = \bigcup_{E \in \mathcal{A}_p(G)} \operatorname{res}^*_{G,E}\big(V_E(k)\big).$$

One way to view a maximal ideal $\mathfrak{m} \in V_G(k)$ is to think of it as the kernel of a nonzero homomorphism $H^*(G,k) \xrightarrow{\alpha} \overline{k}$, where $\overline{k}$ is an algebraic closure of k. After all, $H^*(G,k)/\mathfrak{m}$ is a field which is algebraic over k, and hence it can be embedded in $\overline{k}$. From this point of view, Quillen's Theorem says that every such homomorphism α is a composition of the form

$$H^*(G,k) \xrightarrow{\mathrm{res}_{G,E}} H^*(E,k) \xrightarrow{\beta} \overline{k}$$

for some $E \in \mathcal{A}_p(G)$.

As an example consider $G = D_8 = \langle x, y \mid x^2 = y^2 = (xy)^4 = 1\rangle$, the dihedral group of order 8. This group has two maximal elementary abelian 2-subgroups: $E_1 = \langle x, (xy)^2\rangle$ and $E_2 = \langle y, (xy)^2\rangle$. For $\operatorname{char} k = 2$, the cohomology ring (see Theorem 7.8) has the form

$$H^*(G,k) = k[a_1, b_1, c_2]/(a_1 b_1).$$

Hence there are two minimal prime ideals (a_1) and (b_1), and these correspond to the kernels of the restrictions to the maximal elementary abelian 2-subgroups. That is, $H^*(G,k)/(a_1) \cong k[b_1, c_2]$ is a polynomial ring and is embedded by the restriction as a subalgebra of $H^*(E_i,k)$ for one of the E_i's.

Quillen's Theorem has an equivalent formulation by Quillen and Venkov, which says that an element of $H^*(G,k)$ is nilpotent if and only if its restriction to every elementary abelian p-subgroup is nilpotent.

Definition For a kG-module M let $J(M) = J_G(M) \subseteq H^*(G,k)$ be the annihilator of the cohomology of M as before (cf. page 58). Then let $V_G(M)$, the *variety of M*, be the variety of the ideal $J(M)$. That is, $V_G(M) \subseteq V_G(k)$ is the set of all maximal ideals which contain $J(M)$.

The main issue of this section is to develop some of the properties of the variety of a module. First we state a basic lemma.

Lemma 10.2 *Let M be a kG-module, and let $\zeta \in H^n(G,k)$. If $V_G(M) \subseteq V_G(\zeta)$, then $\zeta^m \in J(M)$ for some m, that is, $\zeta \in \operatorname{rad} J(M)$.*

Proof. The point is that the relationship between ideals in a Noetherian ring R and their varieties goes as

$$I \subseteq J \Longrightarrow V(I) \supseteq V(J) \Longrightarrow \operatorname{rad} I \subseteq \operatorname{rad} J.$$

This is basic commutative algebra. That is, if $V(J) \subseteq V(I)$, then since the variety of R/J is $V(J)$ we have $V_{R/J}(I + J/J) = V_{R/J}(0) = V_R(J)$. So $I + J/J$ is in every maximal ideal of R/J, and hence it is in the radical of R/J. The result follows from the fact that the radical of a Noetherian ring is nilpotent. □

Lemma 10.3 *For any module M in ${}_{kG}\mathfrak{mod}$, $V_G(M) = V_G(M^*) = V_G(\Omega^n(M))$ for all n.*

Proof. We need only note that

$$\widehat{\mathrm{Ext}}^*_{kG}(M,M) \cong \widehat{\mathrm{Ext}}^*_{kG}(M^*,M^*) \cong \widehat{\mathrm{Ext}}^*_{kG}(\Omega^n(M),\Omega^n(M))$$

as $H^*(G,k)$-modules. □

Lemma 10.4 *If $L \longrightarrow M \longrightarrow N \longrightarrow \Omega^{-1}(L)$ is a triangle in ${}_{kG}\mathfrak{stmod}$, then $V_G(M) \subseteq V_G(L) \cup V_G(N)$. (And by triangle shifting it then follows that also $V_G(L) \subseteq V_G(N) \cup V_G(M)$ and $V_G(N) \subseteq V_G(M) \cup V_G(L)$.)*

Proof. We have a long exact sequence

$$\cdots \longrightarrow \widehat{\mathrm{Ext}}^{m-1}_{kG}(L,\quad) \longrightarrow \widehat{\mathrm{Ext}}^{m}_{kG}(N,\quad) \longrightarrow \widehat{\mathrm{Ext}}^{m}_{kG}(M,\quad) \longrightarrow \widehat{\mathrm{Ext}}^{m}_{kG}(L,\quad) \longrightarrow \cdots$$

From this we see that

$$J(M) \supseteq J(L) \cdot J(N) \supseteq \big(J(L) \cap J(N)\big)^2.$$

□

Proposition 10.5 $V_G(M) = \{0\}$ *if and only if M is projective.*

Proof. If M is projective, then surely $V_G(M) = \{0\}$.

So suppose that $V_G(M) = \{0\}$. Let S be any module in ${}_{kG}\mathfrak{mod}$. Because $\mathrm{Ext}^*_{kG}(M,S)$ is finitely generated as an $H^*(G,k)$-module and $H^*(G,k)$ is finitely generated as a k-algebra, $\mathrm{Ext}^m_{kG}(M,S)$ is zero for m sufficiently large. Since there are only finitely many isomorphism classes of irreducible kG-modules, it follows that there is some m such that $\mathrm{Ext}^m_{kG}(M,S) = 0$ for all irreducible modules S.

Now consider $\Omega^m(M)$. If $\Omega^m(M) \neq 0$, then there is a nonzero homomorphism $\varphi : \Omega^m(M) \longrightarrow S$ for some irreducible module S. By the assumption φ factors through a projective. So if P is a projective cover of S, then we have a commutative diagram

$$\begin{array}{ccccc} & & \Omega^m(M) & & \\ & {\scriptstyle\theta}\swarrow & \downarrow{\scriptstyle\varphi} & & \\ P & \xrightarrow{\ \rho\ } & S & \longrightarrow & 0 \end{array}$$

where $\rho\theta = \varphi$. But because $P \overset{\rho}{\twoheadrightarrow} S$ is a projective cover, θ must be surjective (by Nakayama's Lemma). Hence $\Omega^m(M) \xrightarrow{\theta} P \longrightarrow 0$ splits, so that we have a homomorphism $\sigma : P \longrightarrow \Omega^m(M)$ with $\theta\sigma = \mathrm{id}_P$. This implies that $\sigma(P)$ is a nonzero projective direct summand of $\Omega^m(M)$. This is absurd. Hence $\Omega^m(M) = 0$. So also $\Omega^0(M) = \Omega^{-m}(\Omega^m(M)) = 0$, which shows that M is projective. □

Theorem 10.6 *Let M, N be in ${}_{kG}\mathfrak{mod}$. Then*

$$V_G(M \otimes N) = V_G(M) \cap V_G(N).$$

The proof that we give here assumes that the field k is algebraically closed. In a more general situation suitable modifications can be made.

Proof. We have homomorphisms

$$H^*(G, M) \otimes H^*(G, N) \to H^*(G \times G, M \otimes N) \xrightarrow{\mathrm{res}_{G\times G, G}} H^*(G, M \otimes N)$$

which can be used to define the cup product structure on cohomology. Note that in the term $H^*(G \times G, M \otimes N)$ we are regarding $M \otimes N$ as a $k(G \times G)$-module via the isomorphism $k(G \times G) \cong kG \otimes kG$. For $M \cong N \cong k$ we have an induced map on varieties

$$V_G(k) \xrightarrow{\mathrm{res}^*} V_{G\times G}(k)$$

induced by the restriction $H^*(G \times G, k) \xrightarrow{\mathrm{res}} H^*(G, k)$. Recall also that we are thinking of G as the subgroup $\Delta G = \{(g, g) \mid g \in G\}$ of $G \times G$. It follows from all of this that

$$V_G(M \otimes N) = (\mathrm{res}^*)^{-1}\big(V_{G\times G}(M \otimes N)\big).$$

We need to establish two facts.

(1) $J_{G\times G}(M \otimes N) \cong J_G(M) \otimes H^*(G, k) + H^*(G, k) \otimes J_G(N)$. Here $J_{G\times G}(M \otimes N)$ is the annihilator of $\mathrm{Ext}^*_{k(G\times G)}(M \otimes N, M \otimes N)$ in the cohomology ring $H^*(G \times G, k) \cong H^*(G, k) \otimes H^*(G, k)$.

Proof. Let $\sigma_M : H^*(G, k) \to \mathrm{Ext}^*_{kG}(M, M)$ be given by $\sigma_M(\zeta) = \zeta \cdot \mathrm{id}_M$. Then there is an exact sequence

$$0 \longrightarrow J_G(M) \xrightarrow{\gamma_M} H^*(G, k) \xrightarrow{\sigma'_M} \mathrm{im}\,\sigma_M \longrightarrow 0,$$

where $\sigma_M = \big(\mathrm{im}\,\sigma_M \hookrightarrow \mathrm{Ext}^*_{kG}(M, M)\big) \circ \sigma'_M$. Likewise we have an exact sequence

$$0 \longrightarrow J_G(N) \xrightarrow{\gamma_N} H^*(G, k) \xrightarrow{\sigma'_N} \mathrm{im}\,\sigma_N \longrightarrow 0.$$

Now tensor the complexes $\big(J_G(M) \xrightarrow{\gamma_M} H^*(G, k)\big)$ and $\big(J_G(N) \xrightarrow{\gamma_N} H^*(G, k)\big)$ to get an exact sequence

$$0 \to J_G(M) \otimes J_G(N) \to J_G(M) \otimes H^*(G, k) \oplus H^*(G, k) \otimes J_G(N)$$
$$\to H^*(G, k) \otimes H^*(G, k) \xrightarrow{\sigma'_M \otimes \sigma'_N} (\mathrm{im}\,\sigma_M) \otimes (\mathrm{im}\,\sigma_N) \to 0.$$

But $(\mathrm{im}\,\sigma_M) \otimes (\mathrm{im}\,\sigma_N) = H^*(G \times G, k) \cdot \mathrm{id}_{M \otimes N}$ in $\mathrm{Ext}^*_{k(G\times G)}(M \otimes N, M \otimes N)$. So the kernel of $\sigma'_M \otimes \sigma'_N$ is $J_{G\times G}(M \otimes N)$ and has the desired form. □

(2) $V_{G\times G}(k) \cong V_G(k) \times V_G(k)$.

Proof. Suppose that $\mathfrak{m}_\alpha, \mathfrak{m}_\beta \in V_G(k)$ are maximal ideals which are the kernels of the homomorphisms $\alpha, \beta : H^*(G,k) \longrightarrow k$. Define $\psi : V_G(k) \times V_G(k) \longrightarrow V_{G\times G}(k)$ by $\psi(\mathfrak{m}_\alpha, \mathfrak{m}_\beta) := \mathfrak{m}_{\alpha\otimes\beta}$, the kernel of the homomorphism

$$H^*(G\times G,k) \cong H^*(G,k)\otimes H^*(G,k) \xrightarrow{\alpha\otimes\beta} k.$$

If $\mathfrak{m}_\gamma \in V_{G\times G}(k)$ is the kernel of γ, then define $\alpha, \beta : H^*(G,k) \longrightarrow k$ to be the compositions of γ with $\mu, \nu : H^*(G,k) \longrightarrow H^*(G\times G,k)$, where $\mu(\zeta) = \zeta\otimes 1$, $\nu(\zeta) = 1\otimes\zeta$. It is easy to check that $\psi(\mathfrak{m}_\alpha, \mathfrak{m}_\beta) = \mathfrak{m}_\gamma$. □

To finish the proof of the theorem we need to notice that $V_{G\times G}(M\otimes N) \cong V_G(M)\times V_G(N)$ by (1). Also the map $V_G(k) \longrightarrow V_G(k)\times V_G(k)$ induced by the restriction of $G\times G$ onto G is the diagonal homomorphism. So

$$V_G(M\otimes N) = (\mathrm{res}^*)^{-1}\big(V_G(M)\times V_G(N)\big) = V_G(M)\cap V_G(N).$$

□

11 Infinitely generated modules

In this section we wish to discuss some of the problems involved with extending some of the results to the category ${}_{kG}\mathfrak{Mod}$ of all left kG-modules. Let ${}_{kG}\mathsf{st}\mathfrak{Mod}$ denote the stable category of all left kG-modules (even infinitely generated ones) modulo projectives. The main thing we want in this section is to show that ${}_{kG}\mathsf{st}\mathfrak{Mod}$ is a triangulated category. We should recall that the triangulated structure of ${}_{kG}\mathsf{st}\mathfrak{mod}$ depended very much on the fact that injective modules are projective and vice versa. This result can be proved for large classes of rings using various sorts of sophisticated machinery. For group algebras, however, we can stick to reasonably elementary methods.

To start we should show that kG is an injective object in ${}_{kG}\mathfrak{Mod}$. The proof of the injectivity of kG in ${}_{kG}\mathfrak{mod}$ used duality and requires some modification to work with infinitely generated modules. It is still true in ${}_{kG}\mathfrak{Mod}$ that the duality functor $(\ \ )^* = \mathrm{Hom}(\ \ ,k)$ is exact. However, for an infinitely generated module M, $M \ncong M^{**}$. Still, there is a homomorphism $\varphi_M : M \longrightarrow M^{**}$ given by $\varphi_M(m)(\lambda) = \lambda(m)$ for $\lambda \in M^*$, $m \in M$. In addition, if $M \xrightarrow{\theta} N$ is a kG-homomorphism, then the diagram

$$\begin{array}{ccc} M & \xrightarrow{\theta} & N \\ \downarrow{\scriptstyle \varphi_M} & & \downarrow{\scriptstyle \varphi_N} \\ M^{**} & \xrightarrow{\theta^{**}} & N^{**} \end{array}$$

commutes. In particular, if $M \xrightarrow{\theta} N$ with N finitely generated (so that φ_N is an isomorphism), then the composition

$$M \xrightarrow{\varphi_M} M^{**} \xrightarrow{\theta^{**}} N^{**} \xrightarrow{\varphi_N^{-1}} N$$

is θ. With these observations we can suitably modify the proof of Theorem 2.6 to get the injectivity of kG.

To get the injectivity of kG, we will need to reduce to the case that G is a p-group. That is accomplished by the following proposition.

Proposition 11.1 *Let $H \subseteq G$ be a Sylow p-subgroup of G. A kG-module M is projective if and only if M_H is projective. Likewise M is injective if and only if M_H is injective.*

Proof. If M is projective, then it is a direct summand of a free module. So M_H is also projective.

Assume now that M_H is projective and that we have a diagram

$$\begin{array}{ccccc} & & M & & \\ & & \downarrow{\scriptstyle \mu} & & \\ A & \xrightarrow{\sigma} & B & \longrightarrow & 0 \end{array}$$

in ${}_{kG}\mathfrak{Mod}$ with exact row. Because M_H is projective, there is a kH-homomorphism $\theta : M \longrightarrow A$ with $\sigma\theta = \mu$. Define $\hat{\theta} : M \longrightarrow A$ by

$$\hat{\theta}(m) := \frac{1}{[G:H]} \sum_{x \in G/H} x\theta(x^{-1}m)$$

where the sum is over a complete set of representatives of the left cosets of H in G. Then $\hat{\theta}$ is a kG-homomorphism and $\sigma\hat{\theta} = \mu$. Hence M is projective.

A very similar argument proves that if M_H is injective, then M is injective.

So suppose that M is injective and that we are given a diagram

$$\begin{array}{ccccc} 0 & \longrightarrow & A & \xrightarrow{\sigma} & B \\ & & \downarrow{\scriptstyle \mu} & & \\ & & M & & \end{array}$$

where A and B are kH-modules, the maps are kH-homomorphisms, and σ is injective. Then we can induce to get a commutative diagram

$$\begin{array}{ccccc} 0 & \longrightarrow & A & \xrightarrow{\sigma} & B \\ & & \downarrow{\scriptstyle \varphi_A} & & \downarrow{\scriptstyle \varphi_B} \\ 0 & \longrightarrow & A{\uparrow^G} & \xrightarrow{\sigma{\uparrow^G}} & B{\uparrow^G} \\ & & \downarrow{\scriptstyle \hat{\mu}} & \swarrow{\scriptstyle \psi} & \\ & & M & & \end{array}$$

Of course $\sigma{\uparrow^G} = 1 \otimes \sigma : kG \otimes_{kH} A \longrightarrow kG \otimes_{kH} B$. The map φ_A is given by $\varphi_A(a) = 1 \otimes a \in kG \otimes_{kH} A$, and φ_B is defined similarly. For $\sum_{x \in G/H} x \otimes a_x \in A{\uparrow^G}$ we define $\hat{\mu}\Big(\sum_{x \in G/H} x \otimes a_x\Big) = \sum_{x \in G/H} x\mu(a_x)$. It is easy to see that $\sigma{\uparrow^G}$ and $\hat{\mu}$ are kG-homomorphisms. Hence the existence of ψ is a consequence of the injectivity of M. It remains to check that $\hat{\mu} \circ \varphi_A = \mu$. Hence the existence of $\psi \circ \varphi_B : B \longrightarrow M$ proves the injectivity of M_H. □

Theorem 11.2 *A kG-module is projective if and only if it is injective.*

Proof. By the last result we may assume that G is a p-group. Suppose that M is an injective kG-module. Let $S := M^G = \operatorname{soc} M$ be the socle of M, which (since G is a p-group) is also the set of G-fixed points of M. Let B be a k-basis for S and let $F := \bigoplus_{b \in B} kG \cdot y_b$ be a free kG-module with kG-basis consisting of the set of symbols $(y_b)_{b \in B}$. Then we have a commutative diagram

$$\begin{array}{ccccc} 0 & \longrightarrow & S & \xrightarrow{\theta} & F \\ & & {\scriptstyle \iota}\downarrow & \swarrow{\scriptstyle \psi} & \\ & & M & & \end{array}$$

where ι is the inclusion, the kG-homomorphism θ is defined by $\theta(b) := \widetilde{G} \cdot y_b$ for all $b \in B$, and ψ exists by the injectivity of M. Here $\widetilde{G} = \sum_{g \in G} g$. Notice that ψ is injective because it is injective on the socle of F. We need only show that ψ is surjective.

Suppose that $m \in M$ and let $W := \operatorname{ann}_{kG}(m)$ be its annihilator in kG. We claim that $m \in \operatorname{im}\psi$. To see this we proceed by (inverse) induction on the dimension of W. Assume that $u \in M$ is in $\operatorname{im}\psi$ whenever $\dim(\operatorname{ann}_{kG}(u)) > \dim W$. Notice that if $\dim W = |G| - 1$, then $W = \operatorname{rad} kG$, $m \in S$, and $m \in \operatorname{im}\psi$. Now let $x \in kG - W$ be an element such that $x + W \subseteq \operatorname{soc}(kG/W)$. Then $xm \in S$ because for any $g \in G$, $(g-1)xm \in W \cdot m = 0$.

Lemma 11.3 *For x and W as above there exists $y \in kG$ such that $xy = \widetilde{G}$ and $W \cdot y = 0$.*

Proof. Let $A = (kG)^n$ be an injective hull of kG/W. So there is an injection

$$\varphi : kG/W \hookrightarrow A = (kG)^n.$$

Let $\varphi(1) = (z_1, \ldots, z_n) \in (kG)^n$. Then $\varphi(x+W) = x \cdot \varphi(1) = (xz_1, \ldots, xz_n) \neq 0$ but $\varphi(x+W) \in \widetilde{G} \cdot A = (k\widetilde{G})^n$. So $xz_i = a\widetilde{G}$ for some i and some $a \in k$, $a \neq 0$. On the other hand, if $w \in W$, then $0 = \varphi(w+W) = w \cdot \varphi(1) = (wz_1, \ldots, wz_n)$ and hence $wz_i = 0$. So let $y = \frac{1}{a} z_i$. □

Returning to the proof of the theorem we have that $xm \in S = \psi(\widetilde{G} \cdot F)$. So $xm = \psi(\widetilde{G} \cdot f)$ for some $f \in F$. Hence $xm = \psi(xyf)$ for some $y \in kG$ as in the lemma. Hence $\operatorname{ann}_{kG}(m - \psi(yf))$ contains both x and W. By induction $m - \psi(yf) \in \operatorname{im}\psi$ and hence $m \in \operatorname{im}\psi$.

Assuming that G is a p-group we have shown that injective modules are free. If M is projective, then M is a direct summand of a free module and hence is isomorphic to a direct summand of an injective. (Choose any injective module with a basis having large enough cardinality.) But direct summands of injective modules are injective. Hence M is injective. □

Notice that as a by-product of the proof of Theorem 11.2 we have that projective modules over finite p-groups are free. We knew this for finitely generated modules.

Theorem 11.4 *The category $_{kG}\mathrm{st}\mathfrak{Mod}$ is triangulated.*

Proof. The proof is the same as in Section 5. □

We end this section with a note about direct limits (colimits). This material is standard (see [W]), and we omit the proofs.

Definition Suppose that

$$M_1 \xrightarrow{\theta_1} M_2 \xrightarrow{\theta_2} M_3 \xrightarrow{\theta_3} \cdots$$

is a system of modules and homomorphisms. Then $\varinjlim M_i$, the *direct limit* or *colimit* of the system, is a module such that there are homomorphisms

$$\varphi_i : M_i \longrightarrow \varinjlim M_i$$

and the following universal property is satisfied: if for a module N and for each i we have a homomorphism $\sigma_i : M_i \longrightarrow N$ such that

$$\begin{array}{ccc} M_i & \xrightarrow{\theta_i} & M_{i+1} \\ & \sigma_i\searrow \quad \swarrow\sigma_{i+1} & \\ & N & \end{array}$$

commutes, then there is a unique homomorphism $\varinjlim M_i \xrightarrow{\sigma} N$ such that for each i

$$\begin{array}{ccc} M_i & \xrightarrow{\varphi_i} & \varinjlim M_i \\ & \sigma_i\searrow \quad \swarrow\sigma & \\ & N & \end{array}$$

commutes.

Proposition 11.5 *The direct limit of a system of modules exists and is unique up to isomorphism.*

Proof. Let $\gamma : \bigoplus_{i=1}^{\infty} M_i \longrightarrow \bigoplus_{i=1}^{\infty} M_i$ be given by $\gamma(m) = m - \theta_i(m)$ whenever $m \in M_i$. Then it can be proved that $\operatorname{coker} \gamma$ satisfies the condition of the definition. □

Proposition 11.6 *Taking direct limits is an exact functor which commutes with tensor products (over k).*

12 Idempotent modules

In this section we prove the existence of some of the idempotent modules in the stable category $_{kG}\mathfrak{stMod}$. The development follows the ideas of [CW] although this is really only a variation on the original theme of Rickard [R].

Suppose that $V \subseteq V_G(k)$ is a nonempty, closed, homogeneous subvariety. Choose nonzero homogeneous elements $\zeta_i \in H^{n_i}(G,k)$, $n_i > 0$, $(i = 1,\dots,n)$ so that $V = V_G(\zeta_1,\dots,\zeta_n)$. Recall that for each i we have a triangle (see page 60)

$$L_{\zeta_i} \longrightarrow \Omega^{n_i}(k) \longrightarrow k \longrightarrow \Omega^{-1}(L_{\zeta_i})$$

so that there is an exact sequence

$$0 \longrightarrow \Omega^{n_i}(k) \xrightarrow{\hat{\zeta}_i} k \oplus (\text{proj}) \longrightarrow \Omega^{-1}(L_{\zeta_i}) \longrightarrow 0.$$

As before let $\mathcal{C}(\zeta_i)_*$ be the complex $\big(\Omega^{n_i}(k) \xrightarrow{\hat{\zeta}_i} k \oplus (\text{proj})\big)$ having homology $H_*\big(\mathcal{C}(\zeta_i)_*\big) = H_0\big(\mathcal{C}(\zeta_i)_*\big) \cong \Omega^{-1}(L_{\zeta_i})$. For each positive integer m and each $i = 1,\dots,n$ there is a chain map

$$\begin{array}{ccccccccc}
\mathcal{C}(\zeta_i^{m+1})_* : & 0 & \longrightarrow & \Omega^{(m+1)n_i}(k) & \xrightarrow{(\zeta_i^{m+1})\hat{}} & k \oplus (\text{proj}) & \longrightarrow & 0 \\
\downarrow & & & \Big\downarrow{\scriptstyle \Omega^{mn_i}(\hat{\zeta}_i)} & & \downarrow & & \\
\mathcal{C}(\zeta_i^{m})_* : & 0 & \longrightarrow & \Omega^{mn_i}(k) & \xrightarrow{(\zeta_i^{m})\hat{}} & k \oplus (\text{proj}) & \longrightarrow & 0
\end{array}$$

such that the homomorphism in degree zero is equivalent to the identity map on k, modulo projective homomorphisms.

Now take the tensor product $\mathcal{C}_*^{(m)} := \mathcal{C}(\zeta_1^m,\dots,\zeta_n^m)_* := \mathcal{C}(\zeta_1^m)_* \otimes \cdots \otimes \mathcal{C}(\zeta_n^m)_*$. Then we have a system of complexes

$$\begin{array}{cccccccc}
\mathcal{C}_*^{(m+1)} : & \cdots & \longrightarrow & \mathcal{C}_1^{(m+1)} & \longrightarrow & \mathcal{C}_0^{(m+1)} & \longrightarrow & 0 \\
\downarrow & & & \downarrow & & \downarrow & & \\
\mathcal{C}_*^{(m)} : & \cdots & \longrightarrow & \mathcal{C}_1^{(m)} & \longrightarrow & \mathcal{C}_0^{(m)} & \longrightarrow & 0
\end{array}$$

given by taking the tensor product of the chain maps.

Recall that $H_*(\mathcal{C}_*^{(m)}) = H_0(\mathcal{C}_*^{(m)}) \cong \bigotimes_{i=1}^{n} \Omega^{-1}(L_{\zeta_i^m})$ and that, as in the proof of Lemma 9.4, $\partial_1^{(m)}(\mathcal{C}_1^{(m)}) \cong U(\zeta_1^m,\dots,\zeta_n^m)$. So there is a collection of exact sequences

$$E(\zeta_1^m,\dots,\zeta_n^m):$$

$$\begin{array}{ccccccc}
0 \longrightarrow & U(\zeta_1^m,\dots,\zeta_n^m) & \longrightarrow & k \oplus (\text{proj}) & \longrightarrow & \bigotimes_{i=1}^{n} \Omega^{-1}(L_{\zeta_i^m}) & \longrightarrow 0 \\
& \Vert & & & & \Vert & \\
& U^{(m)} & & & & L^{(m)} &
\end{array}$$

Now we take the dual of each of these sequences and consider the directed system of exact sequences where the vertical homomorphisms are induced from the chain maps given previously.

$$\begin{array}{ccccccccc}
0 & \longrightarrow & L^{(1)^*} & \longrightarrow & k \oplus (\mathrm{proj}) & \longrightarrow & U^{(1)^*} & \longrightarrow & 0 \\
 & & \downarrow & & \downarrow & & \downarrow & & \\
0 & \longrightarrow & L^{(2)^*} & \longrightarrow & k \oplus (\mathrm{proj}) & \longrightarrow & U^{(2)^*} & \longrightarrow & 0 \\
 & & \downarrow & & \downarrow & & \downarrow & & \\
0 & \longrightarrow & L^{(3)^*} & \longrightarrow & k \oplus (\mathrm{proj}) & \longrightarrow & U^{(3)^*} & \longrightarrow & 0 \\
 & & \downarrow & & \downarrow & & \downarrow & & \\
 & & \vdots & & \vdots & & \vdots & &
\end{array} \qquad (12.1)$$

Notice that the homomorphisms in the middle column are all equivalent to the identity homomorphism on k, modulo maps which factor through projectives.

Since the direct limit is an exact functor we obtain, in the limit, an exact sequence

$$0 \longrightarrow E \longrightarrow k \oplus (\mathrm{proj}) \xrightarrow{\theta} F \longrightarrow 0. \qquad (12.2)$$

It is an easy check that k is a direct summand of the middle term. The rest is projective because the direct limit of a sequence of projective modules is projective. (In general a direct limit of projectives would only be flat, but in the case of kG-modules flat modules are projective.)

Proposition 12.1 *Let M be a module in ${}_{kG}\mathfrak{mod}$ with $V_G(M) \subseteq V = V_G(\zeta_1, \ldots, \zeta_n)$. Then $F \otimes M =$ (proj) and $E \otimes M \cong M \oplus$ (proj), where E and F are as in* (12.2).

Proof. According to the hypothesis, for any large enough integer m, we have that $\zeta_1^m, \ldots, \zeta_n^m \in J(M)$. It follows that for m sufficiently large the maps

$$C_i^{(r+m)} \otimes M \to C_i^{(r)} \otimes M$$

coming from the compositions of the chain maps factor through projectives for all $i > 0$. So there are chain maps of exact sequences

$$\begin{array}{ccccccccc}
0 \longrightarrow & C_n^{(r+m)} \otimes M & \xrightarrow{\partial_n^{(r+m)}} \cdots \longrightarrow & C_1^{(r+m)} \otimes M & \longrightarrow & U^{(r+m)} \otimes M & \longrightarrow 0 \\
 & \downarrow & & \downarrow & & \downarrow{\scriptstyle \psi_0^{(r,m)}} & \\
0 \longrightarrow & C_n^{(r)} \otimes M & \xrightarrow{\partial_n^{(r)}} \cdots \longrightarrow & C_1^{(r)} \otimes M & \longrightarrow & U^{(r)} \otimes M & \longrightarrow 0
\end{array}$$

We now claim that for m sufficiently large, $\psi_0^{(r,m)}$ also factors through a projective. This can be proved by successively applying the following lemma to the sequences

$$\begin{array}{ccccccccc} 0 & \longrightarrow & C_n^{(r)} \otimes M & \longrightarrow & C_{n-1}^{(r)} \otimes M & \longrightarrow & \partial(C_{n-1}^{(r)}) \otimes M & \longrightarrow & 0, \\ 0 & \longrightarrow & \partial(C_{n-1}^{(r)}) \otimes M & \longrightarrow & C_{n-2}^{(r)} \otimes M & \longrightarrow & \partial(C_{n-2}^{(r)}) \otimes M & \longrightarrow & 0, \\ & & & & \vdots & & & & \\ 0 & \longrightarrow & \partial(C_2^{(r)}) \otimes M & \longrightarrow & C_1^{(r)} \otimes M & \longrightarrow & U^{(r)} \otimes M & \longrightarrow & 0. \end{array}$$

Lemma 12.2 *Suppose that we have a commutative diagram with exact rows*

$$\begin{array}{ccccccccc} 0 & \longrightarrow & A & \longrightarrow & B & \longrightarrow & C & \longrightarrow & 0 \\ & & \downarrow{\scriptstyle \gamma_1} & & \downarrow{\scriptstyle \gamma_2} & & \downarrow{\scriptstyle \gamma_3} & & \\ 0 & \longrightarrow & A' & \longrightarrow & B' & \longrightarrow & C' & \longrightarrow & 0 \\ & & \downarrow{\scriptstyle \delta_1} & & \downarrow{\scriptstyle \delta_2} & & \downarrow{\scriptstyle \delta_3} & & \\ 0 & \longrightarrow & A'' & \longrightarrow & B'' & \longrightarrow & C'' & \longrightarrow & 0 \end{array}$$

and that γ_1, γ_2, δ_1, and δ_2 all factor through projectives. Then $\delta_3 \circ \gamma_3$ factors through a projective.

Proof. In the stable category we have a diagram of triangles

$$\begin{array}{ccccccc} A & \longrightarrow & B & \longrightarrow & C & \longrightarrow & \Omega^{-1}(A) \\ \downarrow & & \downarrow{\scriptstyle 0} & & \downarrow{\scriptstyle \hat{\gamma}_3} & & \downarrow{\scriptstyle 0} \\ A' & \longrightarrow & B' & \stackrel{\nu}{\longrightarrow} & C' & \stackrel{\mu}{\longrightarrow} & \Omega^{-1}(A') \\ \downarrow & & \downarrow{\scriptstyle 0} & & \downarrow{\scriptstyle \hat{\delta}_3} & & \downarrow{\scriptstyle 0} \\ A'' & \longrightarrow & B'' & \longrightarrow & C'' & \longrightarrow & \Omega^{-1}(A'') \end{array}$$

Because $\mu \circ \hat{\gamma}_3 = 0$, there must exist $\sigma : C \longrightarrow B'$ such that $\hat{\gamma}_3 = \nu\sigma$. But then by the commutativity $\hat{\delta}_3 \circ \hat{\gamma}_3 = \hat{\delta}_3 \circ \nu \circ \sigma = 0$ in the stable category. □

Returning to the proof of Proposition 12.1 we consider the directed system (12.1) $\otimes M$. In the right hand column we have the system

$$U^{(1)^*} \otimes M \to U^{(2)^*} \otimes M \to \cdots$$

which has limit $\varinjlim(U^{(i)^*} \otimes M) = F \otimes M$. But any sufficiently long composition of the homomorphisms must factor through a projective. From this we conclude that $F \otimes M$ is projective. So the sequence

$$0 \longrightarrow E \otimes M \longrightarrow \big(k \oplus (\text{proj})\big) \otimes M \longrightarrow F \otimes M \longrightarrow 0$$

is split and $E \otimes M \cong M \oplus (\text{proj})$. □

Definition Let $\mathfrak{M}_V$ denote the full subcategory of ${}_{kG}\mathfrak{stmod}$ consisting of all (finitely generated, left) kG-modules M with $V_G(M) \subseteq V$. Let $E(V) = E$ and $F(V) = F$ for E and F in the sequence (12.2), where $V = V_G(\zeta_1, \ldots, \zeta_n)$.

We say that an object X in ${}_{kG}\mathfrak{stMod}$ is $\mathfrak{M}_V$*-local* if $\underline{\mathrm{Hom}}_{kG}(M, X) = 0$ for all M in $\mathfrak{M}_V$.

Proposition 12.3 *Suppose that X is $\mathfrak{M}_V$-local and that $\varphi : k \longrightarrow X$ is a homomorphism. Then in the stable category φ is equivalent to $\mu\theta$ for some homomorphism $\mu : F(V) \longrightarrow X$.*

In other words, the homomorphism $\theta : k \longrightarrow F(V)$ is universal for maps from k to $\mathfrak{M}_V$-local objects. We should notice that $F(V)$ is itself an $\mathfrak{M}_V$-local object since $\underline{\mathrm{Hom}}_{kG}\big(M, F(V)\big) \cong \underline{\mathrm{Hom}}_{kG}\big(k, M^* \otimes F(V)\big) = 0$. This is true because M is finitely generated and $V_G(M^*) = V_G(M) \subseteq V$.

Proof. We know that $E(V) = \varinjlim L^{(i)^*}$ is the direct limit of the system

$$L^{(1)^*} \xrightarrow{\ \sigma_1\ } L^{(2)^*} \xrightarrow{\ \sigma_2\ } \cdots .$$

So we have an exact sequence

$$0 \longrightarrow \bigoplus_{i=1}^{\infty} L^{(i)^*} \xrightarrow{\ \sigma\ } \bigoplus_{i=1}^{\infty} L^{(i)^*} \longrightarrow E(V) \longrightarrow 0$$

where $\sigma(l) = l - \sigma_i(l)$ for $l \in L^{(i)^*}$. The fact that $E(V)$ is isomorphic to the cokernel of σ is the essence of the proof of the existence of direct limits in the last section. So the exactness of the above sequence can be established by verifying that σ is injective. This is a straightforward exercise. Now we do a triangle shift on the sequence to get an exact sequence

$$0 \longrightarrow \bigoplus_{i=1}^{\infty} L^{(i)^*} \longrightarrow E(V) \oplus (\text{proj}) \longrightarrow \bigoplus_{i=1}^{\infty} \Omega^{-1}(L^{(i)^*}) \longrightarrow 0.$$

Then we have a long exact sequence

$$\to \underline{\mathrm{Hom}}_{kG}\Big(\bigoplus_{i=1}^{\infty} \Omega^{-1}(L^{(i)^*}), X\Big) \to \underline{\mathrm{Hom}}_{kG}\big(E(V), X\big) \to \underline{\mathrm{Hom}}_{kG}\Big(\bigoplus_{i=1}^{\infty} L^{(i)^*}, X\Big) \to .$$

Now we claim that $\underline{\mathrm{Hom}}_{kG}(\bigoplus_{i=1}^{\infty} L^{(i)^*}, X) = 0$. For suppose that $f : \bigoplus_{i=1}^{\infty} L^{(i)^*} \longrightarrow X$ is a kG-homomorphism. Then let $f_i : L^{(i)^*} \longrightarrow X$ be the restriction of f to $L^{(i)^*}$. Because X is $\mathfrak{M}_V$-local, $f_i = \beta_i \circ \alpha_i$ for some maps $\alpha_i : L^{(i)^*} \longrightarrow P_i$, $\beta_i : P_i \longrightarrow X$, where P_i is some projective kG-module. So let $\alpha : \bigoplus_{i=1}^{\infty} L^{(i)^*} \longrightarrow \bigoplus_{i=1}^{\infty} P_i$ be defined by α_i on $L^{(i)^*}$. Let $\beta : \bigoplus_{i=1}^{\infty} P_i \longrightarrow X$ be given by $\beta(g) = \beta_i(g)$ for $g \in P_i$. It is an easy check that $\beta\alpha = f$, thus establishing the claim.

Similarly, it can be shown that $\underline{\mathrm{Hom}}_{kG}(\bigoplus_{i=1}^{\infty} \Omega^{-1}(L^{(i)^*}), X) = 0$ and hence $\underline{\mathrm{Hom}}_{kG}(E(V), X) = 0$ by the exact sequence. Now by the exact sequence (12.2) it can be seen that

$$\underline{\mathrm{Hom}}_{kG}(F(V), X) \xrightarrow{\ \theta^*\ } \underline{\mathrm{Hom}}_{kG}(k, X)$$

is surjective. □

Proposition 12.4 *In the stable category ${}_{kG}\mathbf{st}\mathfrak{Mod}$ we have that $E(V) \otimes X \cong 0$ for any $\mathfrak{M}_V$-local object X. In particular, $E(V) \otimes F(V) \cong 0$ and by consequence $E(V) \otimes E(V) \cong E(V)$ and $F(V) \otimes F(V) \cong F(V)$.*

Proof. By assumption on X, $\underline{\mathrm{Hom}}_{kG}(L^{(i)}, X) = 0$ for any i. Moreover

$$\underline{\mathrm{Hom}}_{kG}(L^{(i)} \otimes S, X) \cong \underline{\mathrm{Hom}}_{kG}(S, L^{(i)^*} \otimes X) = 0$$

for any finitely generated kG-module S. It follows that $L^{(i)^*} \otimes X$ is projective. But $E(V) \otimes X$ is isomorphic to the direct limit of the system

$$L^{(1)^*} \otimes X \longrightarrow L^{(2)^*} \otimes X \longrightarrow \cdots.$$

Hence $E(V) \otimes X$ is projective. Furthermore, $E(V) \otimes F(V) \cong 0$ because $F(V)$ is $\mathfrak{M}_V$-local. The last statement is a consequence of taking the tensor products of $E(V)$ and $F(V)$ with the exact sequence $0 \to E(V) \to k \oplus (\mathrm{proj}) \to F(V) \to 0$. □

Remarks

(1) The results in this direction, by Rickard [R], are much stronger than those that we have proved in these notes. He shows that for any thick subcategory $\mathfrak{M}$ of ${}_{kG}\mathbf{stmod}$ ("thick" means triangulated and closed under the taking of direct summands) and any object X of ${}_{kG}\mathbf{stmod}$, there is a distinguished triangle

$$\mathcal{E}_{\mathfrak{M}}(X) \longrightarrow X \longrightarrow \mathcal{F}_{\mathfrak{M}}(X) \longrightarrow \Omega^{-1}(\mathcal{E}_{\mathfrak{M}}(X))$$

such that $\mathcal{E}_{\mathfrak{M}}(X)$ is a filtered colimit of objects of $\mathfrak{M}$ and $\mathcal{F}_{\mathfrak{M}}(X)$ is $\mathfrak{M}$-local. The conditions satisfied by $E_{\mathfrak{M}}(X)$ and $F_{\mathfrak{M}}(X)$ actually characterize the triangle. We get some glimpse of this in the theorem that follows.

(2) The relations in Proposition 12.4 are what have led us to call $E(V)$ and $F(V)$ the idempotent modules associated to V. We should note that by Theorem 3.5 the trivial module is the only nonzero object in ${}_{kG}\mathfrak{stmod}$ with this idempotent property.

(3) A direct limit as we have defined it, is a colimit in the literature. A "filtered colimit" is a more general type of colimit which, for example, can be indexed by certain partially ordered sets, not just the natural numbers. While the concept is convenient for many applications we can avoid it in our considerations.

Theorem 12.5 *Let $V \subseteq V_G(k)$ be a nonempty, closed, homogeneous subvariety. The triangle*

$$E(V) \longrightarrow k \longrightarrow F(V) \longrightarrow \Omega^{-1}(E(V))$$

is characterized by the properties:

(1) *The middle term is k,*

(2) *$E(V)$ is a direct limit (filtered colimit) of objects in $\mathfrak{M}_V$,*

(3) *$F(V)$ is $\mathfrak{M}_V$-local.*

Proof. (Sketch) Suppose that $E' \to k \to F' \to \Omega^{-1}(E')$ is another such triangle. If we add suitable projectives, then we can construct the following commutative diagram with exact rows and columns.

$$\begin{array}{ccccccccc}
 & & & & 0 & & 0 & & \\
 & & & & \downarrow & & \downarrow & & \\
 & & & & W & = & W & & \\
 & & & & \downarrow & & \downarrow & & \\
0 & \longrightarrow & k & \longrightarrow & F(V) \oplus (\text{proj}) & \longrightarrow & \Omega^{-1}(E(V)) \oplus (\text{proj}) & \longrightarrow & 0 \\
 & & \| & & \downarrow{\scriptstyle\theta} & & \downarrow{\scriptstyle\varphi} & & \\
0 & \longrightarrow & k & \longrightarrow & F' \oplus (\text{proj}) & \longrightarrow & \Omega^{-1}(E') & \longrightarrow & 0 \\
 & & & & \downarrow & & \downarrow & & \\
 & & & & 0 & & 0 & &
\end{array}$$

Note that θ exists by Proposition 12.3 since F' is $\mathfrak{M}_V$-local. Then φ is the induced map. The point now is to show that the kernel, W, is both $\mathfrak{M}_V$-local (by a suitable long exact sequence of the middle columns) and a direct limit of objects in $\mathfrak{M}_V$ (by the last column). Hence we must have that $\underline{\mathrm{Hom}}_{kG}(W, W) = 0$. So W is projective. □

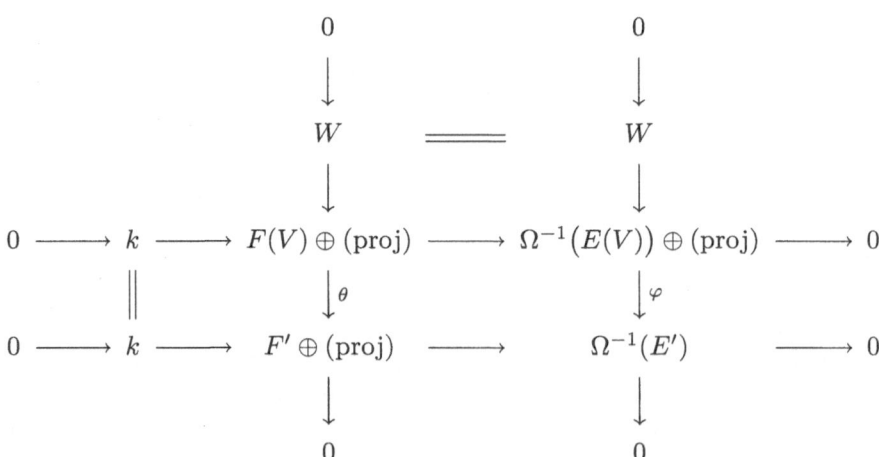

For an easy application consider the following corollary.

Corollary 12.6 *Suppose that V_1 and V_2 are nonempty, closed, homogeneous subvarieties of $V_G(k)$. Then in ${}_{kG}\mathfrak{stMod}$*

$$E(V_1) \otimes E(V_2) \cong E(V_1 \cap V_2).$$

Proof. Consider the complexes $\big(k \oplus (\text{proj}) \xrightarrow{\theta_1} F(V_1)\big)$ and $\big(k \oplus (\text{proj}) \xrightarrow{\theta_2} F(V_2)\big)$ where θ_1 and θ_2 are surjective. Then taking the tensor product we get an exact sequence

$$0 \to E(V_1) \otimes E(V_2) \xrightarrow{\sigma_1 \otimes \sigma_2} k \oplus (\text{proj}) \to F(V_1) \oplus F(V_2) \oplus (\text{proj}) \to F(V_1) \otimes F(V_2) \to 0.$$

Let $F' := \text{coker}(\sigma_1 \otimes \sigma_2)$. Then from the triangle

$$F' \longrightarrow F(V_1) \oplus F(V_2) \longrightarrow F(V_1) \otimes F(V_2) \longrightarrow \Omega^{-1}(F')$$

we get that F' is $\mathfrak{M}_{V_1 \cap V_2}$-local. But also $E(V_1) \otimes E(V_2)$ is a direct limit of objects in $\mathfrak{M}_{V_1 \cap V_2}$. So the triangle

$$E(V_1) \otimes E(V_2) \longrightarrow k \longrightarrow F' \longrightarrow \Omega^{-1}\big(E(V_1) \otimes E(V_2)\big)$$

is isomorphic to the triangle

$$E(V_1 \cap V_2) \longrightarrow k \longrightarrow F(V_1 \cap V_2) \longrightarrow \Omega^{-1}\big(E(V_1 \cap V_2)\big).$$

□

13 Varieties and induced modules

We end these notes with a brief sketch of an application of the idempotent module technology. The application is a slight generalization of a theorem due to Dave Benson [B3]. Its proof involves several concepts which have not been previously covered in these notes. We give a brief description of the necessary material.

To begin we recall Quillen's Dimension Theorem (Theorem 10.1) that

$$V_G(k) = \bigcup_{A \in \mathcal{A}_p(G)} \text{res}^*_{G,A}\big(V_A(k)\big)$$

where $\mathcal{A}_p(G)$ is the collection of all elementary abelian p-subgroups of G. We say that a closed subvariety V of $V_G(k)$ is *supported* on $A \in \mathcal{A}_p(G)$ if $V \subseteq \text{res}^*_{G,A}\big(V_A(k)\big)$. By Quillen's Theorem every irreducible subvariety of $V_G(k)$ is supported on some $A \in \mathcal{A}_p(G)$.

Now suppose that N is a normal subgroup of G and that $P_* \xrightarrow{\varepsilon} k$ is a kG-projective resolution of k. Then for each $n \geqslant 0$, $\mathrm{Hom}_{kN}(P_n, k)$ is a kG-module by the action $(gf)(x) = f(g^{-1}x)$ for $g \in G$, $x \in P_n$. Note that N acts trivially. So we may also consider $\mathrm{Hom}_{kN}(P_n, k)$ as a $k(G/N)$-module. Hence the complex $\mathrm{Hom}_{kN}(P_*, k)$ is a complex of $k(G/N)$-modules, and we have an action of G/N on $H^*(N, k)$. It follows that there is also an action of G/N on $V_N(k)$ induced by the action on $H^*(N, k)$. That is, if $\mathfrak{m}$ is a maximal ideal in $H^*(N, k)$, then so also is $g \cdot \mathfrak{m}$ for any $g \in G$. Clearly the action takes closed sets to closed sets and homogeneous subvarieties to homogeneous subvarieties.

With these notions we are prepared to state the application.

Theorem 13.1 *Let A be an elementary abelian p-subgroup of G and $C := C_G(A)$. Suppose that $V \subseteq V_C(k)$ is a homogeneous subvariety which is supported on A. Assume also that*

(i) *for $g \in N_G(A) - C_G(A)$ we have $g(V) \cap V = \{0\}$ and*

(ii) *for $g \notin N_G(A)$ no point of V is supported on $A \cap gCg^{-1}$.*

Let $W := \mathrm{res}^*_{G,C}(V) \subseteq V_G(k)$. If M is any finitely generated, indecomposable kG-module with $V_G(M) \subseteq W$, then for some kC-module L

$$L{\uparrow}^G \cong M \oplus (\mathrm{proj}).$$

Proof. (Sketch) The whole trick is to show that $E(W)$, the idempotent module defined in the last section, is induced from a kC-module. Specifically we can prove that $E_C(V){\uparrow}^G \cong E(W) \oplus (\mathrm{proj})$. Then, by Frobenius reciprocity, if $V_G(M) \subseteq W$, then

$$M \oplus (\mathrm{proj}) \cong M \otimes \big(E(W) \oplus (\mathrm{proj})\big) \cong \big(M_C \otimes E_C(V)\big){\uparrow}^G.$$

So suppose that L is the nonprojective part of $M_C \otimes E_C(V)$. Then L is a direct summand of $L{\uparrow}^G{\downarrow}_C$, and hence L is indecomposable, finitely generated and is the nonprojective part of $M{\downarrow}_C$. Clearly $L{\uparrow}^G \cong M \oplus (\mathrm{proj})$.

So it remains to show that $E_C(V){\uparrow}^G \cong E(W) \oplus (\mathrm{proj})$. For this we use the characterization of the triangles in Theorem 12.5. First consider the triangle

$$E_C(V) \xrightarrow{\ \sigma\ } k \longrightarrow F_C(V) \longrightarrow \Omega^{-1}(E_C(V)).$$

In ${}_{kC}\mathfrak{Mod}$ we have a homomorphism $\sigma' : E_C(V) \to k$ which represents σ and in ${}_{kG}\mathfrak{Mod}$ we get homomorphisms

$$\big(E_C(V)\big){\uparrow}^G \xrightarrow{\ \hat{\sigma}\ } k_C{\uparrow}^G \xrightarrow{\ \varepsilon\ } k.$$

Here ε is the usual augmentation map, taking $\sum_{x\in G/C} x \otimes a_x \in kG \otimes_{kC} k$ to the sum $\sum_{x\in G/C} a_x \in k$, and $\hat{\sigma}$ is the induced homomorphism. Now we complete $\varepsilon \circ \hat{\sigma}$ to a triangle

$$E_C(V){\uparrow}^G \xrightarrow{\varepsilon\circ\hat{\sigma}} k \longrightarrow F' \longrightarrow \Omega^{-1}(E_C(V){\uparrow}^G). \qquad (13.1)$$

Our aim is to show that the triangle (13.1) is equivalent to the triangle for $E(W)$. To do so we need only verify two things.

(A) $E_C(V){\uparrow}^G$ is a direct limit of finitely generated modules whose varieties are all in W.

This point is relatively easy because $E_C(V)$ is the direct limit of the system

$$L_1 \longrightarrow L_2 \longrightarrow \cdots$$

of kC-modules with $V_C(L_i) \subseteq V$. Then $E_C(V){\uparrow}^G$ is the direct limit of the system $L_1{\uparrow}^G \to L_2{\uparrow}^G \to \cdots$. By the Eckmann-Shapiro Lemma for Ext

$$\mathrm{Ext}^*_{kG}(\quad, L_i{\uparrow}^G) \cong \mathrm{Ext}^*_{kC}(\quad, L_i),$$

and it can be shown from this that

$$V_G(L_i{\uparrow}^G) = \mathrm{res}^*_{G,C}(V_C(L_i)) \subseteq W.$$

(B) The third object, F', in the triangle of $\varepsilon \circ \hat{\sigma}$ is $\mathfrak{M}_W$-local.

Suppose that $M \in \mathfrak{M}_W$. We want to show that $\underline{\mathrm{Hom}}_{kG}(M, F') \cong \underline{\mathrm{Hom}}_{kG}(k, M^* \otimes F') = 0$. It will certainly be sufficient to show that $M^* \otimes F'$ is projective. At this point we need to appeal to the infinitely-generated-module version of Dade's Lemma in [BCR2]. A brief explanation goes as follows.

First let K be an algebraically closed extension of k of large transcendence degree. For $A \in \mathcal{A}_p(G)$, assume that $A = \langle x_1, \ldots, x_n\rangle \cong (\mathbb{Z}/p)^n$. A *cyclic shifted subgroup* of KA is a subgroup of the units of KA of the form $\langle u_\alpha\rangle$ with $u_\alpha = 1 + \sum_{i=1}^{n} \alpha_i(x_i - 1)$ for $\alpha = (\alpha_1, \ldots, \alpha_n) \in K^n - \{0\}$. Each cyclic shifted subgroup $\langle u_\alpha\rangle$ defines a line $\ell \subseteq V_G(K)$ which consists of all maximal ideals which are the kernels of homomorphisms to K which factor through the restriction to $K\langle u_\alpha\rangle$. The extension of Dade's Lemma to infinitely generated modules says that a kG-module is projective if and only if $(K \otimes M){\downarrow}_{\langle u_\alpha\rangle}$ is a projective $K\langle u_\alpha\rangle$-module for all $\alpha \in K^n$, $\alpha \neq 0$. Chouinard's Theorem says that M is projective if and only if its restriction to A is projective for every $A \in \mathcal{A}_p(G)$.

So the question of the projectivity of $M^* \otimes F'$ is reduced to the question of the projectivity of the restrictions of $K \otimes M^* \otimes F'$ to the cyclic shifted subgroups. If either $K \otimes M^*$ or $K \otimes F'$ is projective as a $K\langle u_\alpha \rangle$-module, then so is $K \otimes M^* \otimes F'$. So it is only necessary to check the projectivity at points α corresponding to lines in V $\big($or $\operatorname{res}^*_{G,C}(V) = W\big)$ because for all other α's, $(K \otimes M)\downarrow_{\langle u_\alpha \rangle}$ is projective as $V_G(M) \subseteq W$.

Hence we need only look at the restrictions of F' to $\langle u_\alpha \rangle$ for $\alpha \in V_A(K)$ and $\operatorname{res}^*_{C,A}(\alpha) \in K \otimes V$. Now, by the Mackey formula

$$E_C(V)\uparrow^G\downarrow_A \cong \sum_{x \in A \backslash G / C} (x \otimes E(V))\downarrow_{A \cap xCx^{-1}}\uparrow^A,$$

and the restriction of $\varepsilon \circ \hat{\sigma}$ to each factor is again a composition

$$(x \otimes E(V))\downarrow_{A \cap xCx^{-1}}\uparrow^A \longrightarrow k_{A \cap xCx^{-1}}\uparrow^A \longrightarrow k.$$

Now consider the conditions (i) and (ii) of the theorem. These conditions assure that if $x \notin C$, then $K \otimes \big((x \otimes E(V))\downarrow_{A \cap xCx^{-1}}\uparrow^A\big)$ is free as a $K\langle u_\alpha \rangle$-module. On the other hand if $x \in C$, then we may assume that $x = 1$ and that the triangle of the restriction of $\varepsilon \circ \hat{\sigma}$ to A is just

$$E_A(V') \xrightarrow{\;\tilde{\sigma}\;} k \longrightarrow F_A(V') \longrightarrow \Omega^{-1}\big(E_A(V')\big),$$

the triangle of idempotent modules for $V' = {\operatorname{res}^*_{C,A}}^{-1}(V)$. (Note here that $\operatorname{res}^*_{C,A}$ is injective on $V' \subseteq V_A(k)$.) So $1 \otimes \tilde{\sigma}$ is a $K\langle u_\alpha \rangle$-isomorphism in the stable category. From this we get that

$$K \otimes E(V)\uparrow^G \xrightarrow{1 \otimes (\varepsilon \circ \hat{\sigma})} K$$

is a $K\langle u_\alpha \rangle$-isomorphism in the stable category ${}_{K\langle u_\alpha \rangle}\mathbf{st}\mathfrak{Mod}$. So the third object $K \otimes F'$ is projective as a $K\langle u_\alpha \rangle$-module. □

Remark Benson's original theorem [B3] was stated only for the case that W is a line in $V_G(k)$. That case can actually be proved without appealing to infinitely generated modules. The techniques of [C3] can be easily generalized to show that (with W a line) there exist positive integers l and m and a kG-module L such that there is an exact sequence

$$0 \longrightarrow L \longrightarrow (\Omega^l(k))^m \oplus (\text{proj}) \longrightarrow k_C\uparrow^G \longrightarrow 0$$

where the kernel L has the property that $V_G(L) \cap W = \{0\}$. Hence if $M \in \mathfrak{M}_W$, then $M \otimes L$ is projective. So

$$\begin{aligned} M^m \oplus (\text{proj}) &\cong \Omega^{-l}\big((\Omega^l(k))^m \oplus (\text{proj})\big) \otimes M \oplus (\text{proj}) \\ &\cong \Omega^{-l}\big(k_C\uparrow^G \otimes M\big) \oplus (\text{proj}) \cong \Omega^{-l}\big(M_C\uparrow^G\big) \oplus (\text{proj}) \\ &\cong \big(\Omega^{-l}(M_C)\big)\uparrow^G \oplus (\text{proj}). \end{aligned}$$

So by the Krull-Schmidt Theorem $M \oplus (\mathrm{proj})$ is induced from a kC-module.

Interestingly, this proof does not seem to generalize to give Theorem 13.1. The proof used transfer maps and worked well as long as ordinary cohomology was considered.

As an application of the theorem we consider the example of G being a Sylow 2-subgroup of $\mathrm{SL}(3, 2^n)$, with $p = 2$. The reader should notice that the same idea will work for p odd and G a Sylow p-subgroup of $\mathrm{SL}(3, p^n)$. However, the notation is slightly more complicated.

The field $\mathbb{F}_{2^n}$ can be written as $\mathbb{F}_{2^n} = \mathbb{F}_2(\omega)$ where ω is a primitive $(2^n - 1)^{\mathrm{st}}$ root of 1. Then the set $\{1, \omega, \ldots, \omega^{n-1}\}$ is a basis for $\mathbb{F}_{2^n}$ over the prime field $\mathbb{F}_2$. Let

$$x_i = \begin{bmatrix} 1 & \omega^{i-1} & 0 \\ & 1 & 0 \\ 0 & & 1 \end{bmatrix}, \quad y_i = \begin{bmatrix} 1 & 0 & 0 \\ & 1 & \omega^{i-1} \\ 0 & & 1 \end{bmatrix}, \quad z_i = \begin{bmatrix} 1 & 0 & \omega^{i-1} \\ & 1 & 0 \\ 0 & & 1 \end{bmatrix}.$$

Then G is generated by $x_1, \ldots, x_n, y_1, \ldots, y_n$, and the center and commutator subgroup of G is generated by $z_1, \ldots, z_n$. Let $A := \langle x_1, \ldots, x_n \rangle$, so that $C = C_G(A) = \langle x_1, \ldots, x_n, z_1, \ldots, z_n \rangle$. Both A and C are elementary abelian and $C \lhd G$. Notice also that if $g \in G - C$, then $C_G(g) \cap C = Z(G) = \langle z_1, \ldots, z_n \rangle$. Now $C \cong (\mathbb{Z}/2)^{2n}$ with the vector $(\alpha_1, \ldots, \alpha_n, \beta_1, \ldots, \beta_n)$ corresponding to the element $\prod_{i,j} x_i^{\alpha_i} z_j^{\beta_j}$. The elements of G act linearly on C, by conjugation. For $g \in G$ the matrix of g on C has the form

$$M_g = \begin{bmatrix} \mathrm{id}_n & 0 \\ A_g & \mathrm{id}_n \end{bmatrix}$$

where id_n is the $n \times n$ identity matrix and A_g is an $n \times n$ matrix with coefficients in $\mathbb{F}_2$. The main observation here is the following.

Lemma 13.2 *Suppose that $g \notin C$. Then A_g is nonsingular.*

Now $H^*(C, \mathbb{F}_2) = \mathbb{F}_2[X_1, \ldots, X_n, Z_1, \ldots, Z_n]$ is a polynomial ring in $2n$ variables. The monomials of degree one, $X_1, \ldots, X_n, Z_1, \ldots, Z_n$, form a dual basis to the basis $x_1, \ldots, x_n, z_1, \ldots, z_n$ of C. Furthermore $V_C(k) = k^{2n}$ is the dual space for $k \otimes_{\mathbb{F}_2} H^1(C, \mathbb{F}_2) = H^1(C, k)$. By the double dual theorem $V_C(k) \cong k \otimes_{\mathbb{F}_2} C$, where C is regarded as the vector space $\mathbb{F}_2^{2n}$. Hence the action of $g \in G$ on $V_C(k)$ is also given by the matrix M_g, except of course that we must regard M_g as a matrix with entries in k. Then from the theorem we get the following proposition.

Proposition 13.3

(i) *In the above notation let* $V = \{(\alpha_1, \ldots, \alpha_n, 0, \ldots, 0) \mid \alpha_1, \ldots, \alpha_n \in k\} \subseteq V_C(k)$. *If* M *is a finitely generated* kG*-module with* $V_G(M) \subseteq \mathrm{res}^*_{G,C}(V)$, *then* $M \cong L{\uparrow}^G \oplus (\mathrm{proj})$ *for some* kC*-module* L.

(ii) *Suppose that* $H = \mathrm{SL}(2, 2^n)$. *If* M *is a* kH*-module with* $V_H(M) \subseteq \mathrm{res}^*_{H,C}(V)$, *then* M *is* kC*-projective.*

Proof. The point is that for $g \notin C$, $V \cap g(V) = \{0\}$ by the nonsingularity of the matrix A_g. Hence the result (i) holds because of the theorem. For (ii) we need only notice that M is a direct summand of $M_G{\uparrow}^H$ since $[H : G]$ is prime to p. But by (i) M_G is induced from C. So M is a direct summand of a module induced from C. □

The example can be generalized further to most root subgroups of a Chevalley group in the defining characteristic. But that requires much more notation and analysis.

References

[B1] D. J. Benson: Modular representation theory: new trends and methods. LNM **1081**. Berlin, Heidelberg, New York, Tokyo : Springer 1984

[B2] D. J. Benson: Representations and cohomology. II: Cohomology of groups and modules. Cambridge, New York: Cambridge University Press 1991

[B3] D. J. Benson: Cohomology of modules in the principal block of a finite group. (preprint)

[BC] D. J. Benson and J. F. Carlson: Diagrammatic methods for modular representations and cohomology. Comm. Algebra **15** (1987), 53–121

[BCR1] D. J. Benson, J. F. Carlson, and J. Rickard: Complexity and varieties for infinitely generated modules. Math. Proc. Cambridge Philos. Soc. **118** (1995), 223–243

[BCR2] D. J. Benson, J. F. Carlson, and J. Rickard: Complexity and varieties for infinitely generated modules, II. Math. Proc. Cambridge Philos. Soc. (to appear)

[C1] J. F. Carlson: Module varieties and cohomology rings of finite groups. Vorlesungen aus dem Fachbereich Mathematik der Universität Essen **13** (1985)

[C2] J. F. Carlson: Products and projective resolutions. Proc. Sympos. Pure Math. **47** (1987), 399–408

[C3] J. F. Carlson: Decomposition of the trivial module in the complexity quotient category. J. Pure Appl. Algebra (to appear)

[CDW] J. F. Carlson, P. W. Donovan, and W. W. Wheeler: Complexity and quotient categories for group algebras. J. Pure Appl. Algebra **93** (1994), 147–167

[CP] J. F. Carlson and C. Peng: Relative projectivity and ideals in cohomology rings. J. Algebra (to appear)

[CR1] C. W. Curtis and I. Reiner: Methods of representation theory—with applications to finite groups and orders. Vol. I. New York, Chichester, Brisbane, Toronto: John Wiley & Sons 1981

[CW] J. F. Carlson and W. W. Wheeler: Homomorphisms in higher complexity quotient categories. (preprint)

[E] L. Evens: The cohomology of groups. Oxford Mathematical Monographs. Oxford, New York, Tokyo: Clarendon Press 1991

[H] D. Happel: Triangulated categories in the representation theory of finite-dimensional algebras. Cambridge, New York: Cambridge University Press 1988

[HS] P. J. Hilton and U. Stammbach: A course in homological algebra. GTM 4. New York, Heidelberg, Berlin: Springer 1971

[O] T. Okuyama: A generalization of projective covers of modules over finite group algebras. (unpublished manuscript)

[R] J. Rickard: Idempotent modules in the stable category. (preprint)

[W] C. A. Weibel: An introduction to homological algebra. Cambridge, New York: Cambridge University Press 1994

List of symbols

Index

LM –
Lectures in Mathematics, ETH Zürich

Department of Mathematics
Research Institute of Mathematics

Each year the Eidgenössische Technische Hochschule (ETH) at Zürich invites a selected group of mathematicians to give postgraduate seminars in various areas of pure and applied mathematics. These seminars are directed to an audience of many levels and backgrounds. Now some of the most successful lectures are being published for a wider audience through the ***Lectures in Mathematics, ETH Zürich*** *series. Lively and informal in style, moderate in size and price, these books will appeal to professionals and students alike, bringing a quick understanding of some important areas of current research.*

C. de Boor
Splinefunktionen
1990. ISBN 3-7643-2514-3

J.D. Monk
Cardinal Functions on Boolean Algebras
1990. ISBN 3-7643-2495-3

D. Bättig/H. Knörrer
Singularitäten
1991. ISBN 3-7643-2616-6

R.J. LeVeque
Numerical Methods for Conservation Laws
2nd Edition, 3rd Printing 1994.
1992. ISBN 3-7643-2723-5

R. Narasimhan
Compact Riemann Surfaces
1992. ISBN 3-7643-2742-1

A.J. Tromba
Teichmüller Theory in Riemannian Geometry
1992. ISBN 3-7643-2735-9

M. Yor
Some Aspects of Brownian Motion
1992. ISBN 3-7643-2807-X

G. Baumslag
Topics in Combinatorial Group Theory
1993. ISBN 3-7643-2921-1

M. Giaquinta
Introduction to Regularity Theory for Nonlinear Elliptic Systems
1993. ISBN 3-7643-2879-7

O. Nevanlinna
Convergence of Iterations for Linear Equations
1993. ISBN 3-7643-2865-7

R.-P. Holzapfel
The Ball and Some Hilbert Problems
1995. ISBN 3-7643-2835-5

forthcoming
M. Freidlin
Markov Processes and Differential Equations:
Asymptotic Problems
1996. ISBN 3-7643-5392-9